BASIC ENGINEERING CHEMISTRY

BASIC ENGINEERING CHEMISTRY

S K Singh

Professor

Department of Applied Sciences and Humanities

Institute of Engineering and Technology (IET)

Lucknow

Publishing Globally

NEW AGE INTERNATIONAL (P) LIMITED, PUBLISHERS

LONDON • NEW DELHI • NAIROBI

Bangalore • Chennai • Cochin • Guwahati • Hyderabad • Kolkata • Lucknow • Mumbai

Visit us at **www.newagepublishers.com**

Published by New Age International (P) Ltd., Publishers
First Edition: 2009
Reprint: 2018

GLOBAL OFFICES

- **New Delhi** **NEW AGE INTERNATIONAL (P) LIMITED, PUBLISHERS**
7/30 A, Daryaganj, New Delhi-110002, (INDIA)
Tel.: (011) 23253771, 23253472, **Telefax:** 23267437, 43551305
E-mail: contactus@newagepublishers.com • Visit us at www.newagepublishers.com
- **London** **NEW AGE INTERNATIONAL (UK) LTD.**
27 Old Gloucester Street, London, WC1N 3AX, UK
E-mail: info@newacademicscience.co.uk • Visit us at www.newacademicscience.co.uk
- **Nairobi** **NEW AGE GOLDEN (EAST AFRICA) LTD.**
Ground Floor, Westlands Arcade, Chiromo Road (Next to Naivas Supermarket)
Westlands, Nairobi, KENYA, **Tel.:** 00-254-713848772, 00-254-725700286
E-mail: kenya@newagepublishers.com

BRANCHES

- **Bangalore** 37/10, 8th Cross (Near Hanuman Temple), Azad Nagar, Chamarajpet, Bangalore- 560 018
Tel.: (080) 26756823, **Telefax:** 26756820, **E-mail: bangalore@newagepublishers.com**
- **Chennai** 26, Damodaran Street, T. Nagar, Chennai-600 017, **Tel.:** (044) 24353401, **Telefax:** 24351463
E-mail: chennai@newagepublishers.com
- **Cochin** CC-39/1016, Carrier Station Road, Ernakulam South, Cochin-682 016, **Tel.:** (0484) 2377303, **Telefax:** 4051304
E-mail: cochin@newagepublishers.com
- **Guwahati** Hemsen Complex, Mohd. Shah Road, Paltan Bazar, Near Starline Hotel, Guwahati-781 008,
Tel.: (0361) 2513881, **Telefax:** 2543669, **E-mail: guwahati@newagepublishers.com**
- **Hyderabad** 105, 1st Floor, Madhiray Kaveri Tower, 3-2-19, Azam Jahi Road, Near Kumar Theater, Nimboliadda Kachiguda, Hyderabad-500 027, **Tel.:** (040) 24652456, **Telefax:** 24652457
E-mail: hyderabad@newagepublishers.com
- **Kolkata** RDB Chambers (Formerly Lotus Cinema) 106A, 1st Floor, S N Banerjee Road, Kolkata-700 014
Tel.: (033) 22273773, **Telefax:** 22275247, **E-mail: kolkata@newagepublishers.com**
- **Lucknow** 16-A, Jopling Road, Lucknow-226 001, **Tel.:** (0522) 2209578, 4045297, **Telefax:** 2204098
E-mail: lucknow@newagepublishers.com
- **Mumbai** 142C, Victor House, Ground Floor, N.M. Joshi Marg, Lower Parel, Mumbai-400 013
Tel.: (022) 24927869, **Telefax:** 24915415, **E-mail: mumbai@newagepublishers.com**
- **New Delhi** 22, Golden House, Daryaganj, New Delhi-110 002, **Tel.:** (011) 23262368, 23262370, **Telefax:** 43551305
E-mail: sales@newagepublishers.com

ISBN: 978-81-224-2569-7
C-17-08-10674

Printed in India at Ajit Printing Press, Delhi.
Typeset at Kalyani Computer Services, Delhi.

NEW AGE INTERNATIONAL (P) LIMITED, PUBLISHERS
7/30 A, Daryaganj, New Delhi-110002
Visit us at **www.newagepublishers.com**
(CIN: U74899DL1966PTC004618)

PREFACE

This book is primarily intended to be a textbook for B.E./B.Tech. first year students of Engineering and Technology. This book has a special role to play in the curriculum of all branches of Engineering and Technology and will hopefully present valuable guidelines to young graduates. Efforts have been made to cover the complete syllabus of Engineering Chemistry and Applied Chemistry for B.E./B.Tech. students especially as prescribed by Maharashtra Technical University, Maharashtra. The students as well as engineers in industry in specialized training of different technology *e.g.*, Polymer Technology, Corrosion and its Prevention Technology, Battery Technology, Air and Water Pollution, Catalysis, Adhesives, Lubrication Technology, Colloidal Chemistry, Metal Finishing Technology and Ceramic Technology, etc., will also find this book useful.

A fresh engineer is not only expected to know the basic principles of engineering chemistry but also to apply the knowledge to technology and experimental conditions. An attempt has been made to bridge the gap between the fundamental theory on one hand and the practical use of knowledge in this field on the other hand.

This textbook is the outcome of many years of teaching experience and is based on the lecture notes as well as discussions with various eminent personalities serving in this field for long. A larger number of books and matters available on internet have also been consulted.

This book is divided into 10 chapters, each of which is reasonably complete in itself. Plenty of problems are also provided at the end of each chapter.

We will greatly appreciate being informed of any error in the book. We would also welcome receiving constructive criticism which could be meaningfully utilized in the revision of the book.

S.K. SINGH

ACKNOWLEDGEMENTS

With great esteem and gratitude, I take this opportunity to express my sincere thanks to Prof. G.N. Pandey, Ex. Hon. Vice Chancellor, JGSV University, Chitrakoot, Ex. Director, IET, Lucknow and BIT, Meerut and currently the Faculty at HIT, Allahabad, for his guidance, ever inspiring help and affectionate encouragement during the entire period of the writing of this book.

I am extremely thankful to Prof. D.N. Dhar, Ex. Prof. IIT, Kanpur, and Ex. Director, SIET, Lucknow, Prof. D.C. Rupainwar, Ex. Prof. IT, B.H.U, Varanasi, Dr. S.N. Chauhan Director, SDCET, Muzaffarnagar for their valuable suggestions and encouragement.

I record my indebtedness to Director, Institute of Engineering and Technology, Lucknow, Prof. B. Kunwar, Head, Department of Applied Sciences and Humanities, IET, Lucknow and Prof. V.K. Singh, Controller of Examination, UP Technical University, Lucknow for their contributions and inspirations made during the completion of this book.

I am overwhelmed with gratitude to express my heartful thanks to all my family members whose blessings, inspirations and encouragement, especially the cooperation from my wife Kalpana and daughters Sakshi and Yashi always boosted my determination and enthusiasm for completion of this book.

Last but not the least, I must thank Mr L.N. Mishra, Regional Manager and the editorial department of New Age International (P) Ltd., Publishers for their cooperation, constant moral support and valuable suggestions made during the course of this work.

S.K. SINGH

CONTENTS

CHAPTER

1

ELECTROCHEMISTRY

1.1 INTRODUCTION

Electrochemistry deals with chemical applications of electricity. Substances capable of conducting electricity (conductors) are of two kinds. In metallic, the electrons in the material move under the influence of an applied electric potential–the stream of electrons constituting the current. On the other hand, the passage of electricity through electrolytic conductors or electrolytes is invariably accompanied by chemical decomposition. In order to pass a current of electricity through aqueous solution of electrolytes or through electrolytes in the fused state, two electronic conductors (electrodes) are used. The electrode connected to the positive pole of the current source (battery) is called the anode, while the electrode connected to the negative pole is called the cathode. The flow of electricity is due to the movement of ions of the electrolyte. The ions which carry a positive charge move through the solution towards the cathode and are called **cations.** The negatively charged ions called anions, travel in a opposite direction to the anode. The chemical decomposition of an electrolyte, caused by the passage of electricity, is called **electrolysis.**

Let us consider the electrolysis of an aqueous solution of HCl between platinum electrodes, as shown in figure 1.1. It is seen that hydrogen is evolved at the cathode and chlorine is evolved at the anode on applying a voltage of more than 1.3 V. Since the solution contains hydrogen ions, the evolution of hydrogen can be accounted for by considering that the hydrogen ions gain electrons from the cathode or get reduced at the cathode to give hydrogen, *i.e.*

$$H^+_{(aq)} + e^- \longrightarrow \frac{1}{2} H_2$$

At the anode, the oxidation of chloride ion to give chlorine takes place.

$$Cl^-_{(aq)} - e^- \longrightarrow \frac{1}{2} Cl_2$$

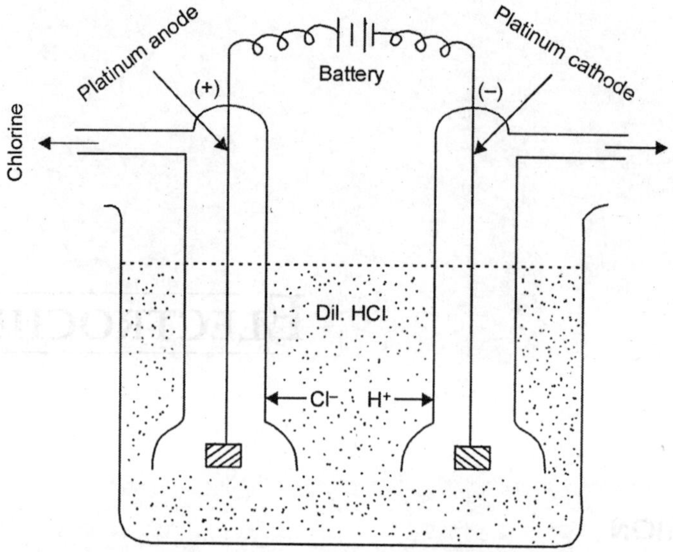

Fig. 1.1: Electrolysis of an aqueous solution of HCl between platinum electrodes

The electrons lost to the anode are 'pumped' by the battery to the cathode so that the reduction of H^+ can take place at the cathode. Thus, *the electrode at which oxidation occurs, is called the anode, and that at which reduction occurs, is called the cathode.*

1.2 ACIDS AND BASES

Boyle in the sixteenth century, defined acids as substances whose aqueous solution

(*i*) turned blue litmus red
(*ii*) neutralised bases
(*iii*) reacted with active metals with evolution of hydrogen
(*iv*) tasted sour.

Bases were similarly defined as substances whose water solution

(*i*) turned red litmus blue
(*ii*) neutralised acids
(*iii*) tasted bitter and
(*iv*) gave soapy touch.

The preceding classification of substances into acids and bases is not satisfactory. We discuss below some of the most commonly used conceptual definitions of acids and bases.

1.2.1 The Arrhenius Theory

According to this theory, "a substance which gives hydrogen ions (H^+) in aqueous solution is called an acid. Similarly, a base is defined as a substance which gives hydroxide ions (OH^-) in aqueous solution" *e.g.*, HNO_3, H_2SO_4 and HCl are called acids because they undergo dissociation in water and yield hydrogen ions.

$$HNO_3 \rightleftharpoons H^+ + NO_3^-$$

$$H_2SO_4 \rightleftharpoons 2H^+ + SO_4^{2-}$$
$$HCl \rightleftharpoons H^+ + Cl^-$$

Compounds such as NaOH, $Ca(OH)_2$ and $Al(OH)_3$ are classified as bases because they yield OH^- ions, when dissolved in water.

$$NaOH \rightleftharpoons Na^+ + OH^-$$
$$Ca(OH)_2 \rightleftharpoons Ca^{2+} + 2OH^-$$
$$Al(OH)_3 \rightleftharpoons Al^{3+} + 3OH^-$$

According to Arrhenius theory, the process of neutralization involves combination of H^+ and OH^- to form water.

$$H^+ + OH^- \rightleftharpoons H_2O$$

Following are shortcomings of the Arrhenious theory:

(*i*) This theory difines acids and bases in terms of aqueous solution and not in the terms of the substances themselves.

(*ii*) This theory defines acids and bases as substances which produce H^+ and OH^- respectively. There are bases none of which contains OH^- ions.

1.2.2 The Bronsted-Lowry Theory

According to this theory, "a substance having a tendency to donate protons is called an acid and a substance having a tendency to accept protons is called a base". In short, acid is a proton donor and base is a proton acceptor. In terms of this definition, an acid A and a base B are related as

$$A \rightleftharpoons B + H^+$$

The species B is called the **conjugate base** of the acid A. Similarly the acid A is called **conjugate acid** of base B. A and B are termed as a conjugate acid-base pair.

In the reaction

$$HCl + H_2O \rightleftharpoons H_3O^+ + Cl^-$$

Hydrogen chloride donates a proton to water and is, therefore, an acid. Water on the other hand, accepts a proton from hydrogen chloride, and is therefore a base in this reaction. In the reverse reaction which at equilibrium proceeds at the same rate as the forward reaction, the hydronium ion (H_3O^+) donates a proton to the chloride ion (Cl^-), hence it is an acid. Chloride ion, because it accepts a proton from the hydronium ion, is a base.

Acid-base pairs such as

$$HCl \underset{+H^+}{\overset{-H^+}{\rightleftharpoons}} Cl^- \quad \text{and}$$

$$H_2O \underset{-H^+}{\overset{+H^+}{\rightleftharpoons}} H_3O^+$$

the members of which can be formed from each other mutually by losing or gaining a proton, are called conjugate acid-base pairs.

The strength of an acid on the Bronsted concept is measured by its tendency to donate a proton; a strong acid is one which has a strong tendency to donate a proton. Similarly, the strength of a base is measured by its tendency to accept a proton; a strong base is one which has a strong tendency to accept a proton *e.g.*,

(*i*) $\underset{\text{strong acid}}{HCl} + \underset{\text{strong base}}{H_2O} \rightleftharpoons \underset{\text{weak acid}}{H_3O^+} + \underset{\text{weak base}}{Cl^-}$

(*ii*) $\underset{\text{strong base}}{NH_2^-} + \underset{\text{strong acid}}{H_2O} \rightleftharpoons \underset{\text{weak acid}}{NH_3} + \underset{\text{weak base}}{OH^-}$

(*iii*) $\underset{\text{weak base}}{NH_3} + \underset{\text{weak acid}}{H_2O} \rightleftharpoons \underset{\text{strong acid}}{NH_4^+} + \underset{\text{strong base}}{OH^-}$

(*iv*) $\underset{\text{weak acid}}{CH_3COOH} + \underset{\text{weak base}}{H_2O} \rightleftharpoons \underset{\text{strong acid}}{H_3O^+} + \underset{\text{strong base}}{CH_3COO^-}$

The reactions considered above reveal that the proton transfer reaction favours the formation of a weaker acid and a weaker base. It may, therefore, be concluded that all proton transfer reactions occur to form predominantly a weaker acid and a weaker base.

According to the Bronsted-Lowry definitions there are three types of acids and bases:

Bronsted-Lowry acids:

Molecular: $HCl \rightarrow H^+ + Cl^-$

Cationic: $[Al(H_2O)_6]^{3+} \rightarrow [Al(H_2O)_5]^{2+} + H^+$

Anionic: $HCO_3^- \rightarrow H^+ + CO_3^{2-}$

Bronsted-Lowry bases:

Molecular: $Py + H^+ \rightarrow [PyH]^+$

Cationic: $[Al(CH_2O)_5OH]^{2+} + H^+ \rightarrow [Al(H_2O)_6]^{3+}$

Anionic: $HCO_3^- + H^+ \rightarrow H_2CO_3$

Amphiprotic Substances

There are many substances which act as an acid in certain reactions and as a base in some other reactions. Such substance are said to be amphiprotic. Water is the most familiar amphoprotic substance. It may accept protons and act as a base as is evident from the following reactions:

$$\underset{\text{Acid}}{HCl} + \underset{\text{Base}}{H_2O} \rightleftharpoons \underset{\text{Acid}}{H_3O^+} + \underset{\text{Base}}{Cl^-}$$

$$CH_3COOH + H_2O \rightleftharpoons H_3O^+ + CH_3COO^-$$

Water molecules may also donate protons and act as an acid.

$$\underset{\text{Acid}}{H_2O} + \underset{\text{Base}}{NH_3} \rightleftharpoons \underset{\text{Acid}}{NH_4^+} + \underset{\text{Base}}{OH^-}$$

$$H_2O + HPO_4^{2-} \rightleftharpoons H_2PO_4^- + OH^-$$

1.2.3 Lewis Concept of Acids and Bases

Lewis defined acid and base in terms of their ability to accept or donate electrons. "Any species capable of donating a pair of electrons to form covalent bond is called a **Lewis base.** Similarly, any species capable of accepting a pair of electrons to form covalent bond is called a **Lewis acid.**" The formation of covalent bond is called **'neutralization'**.

The Lewis acid is also known as **electrophiles** while the Lewis base is known as **nucleophiles** *e.g.,*

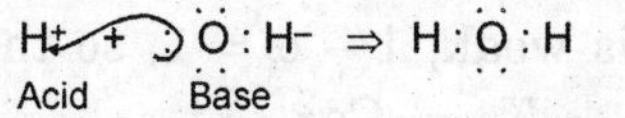

In the above reaction, H^+ is an acid because it accepts a pair of elecrons to form a covalent bond whereas OH^- donates a pair of electrons and is, therefore, a base. Some other examples of acid-base reactions are:

(*i*)

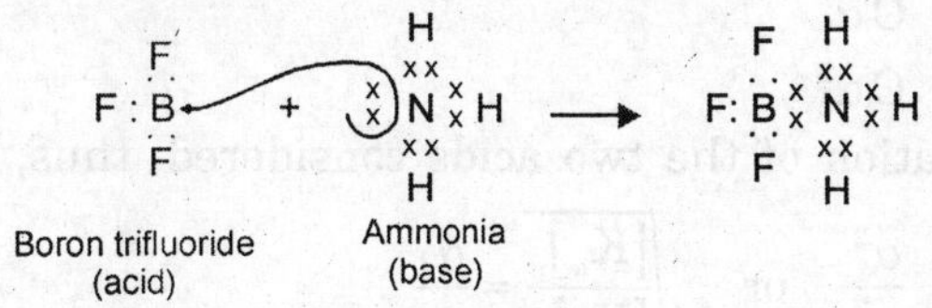

(*ii*)

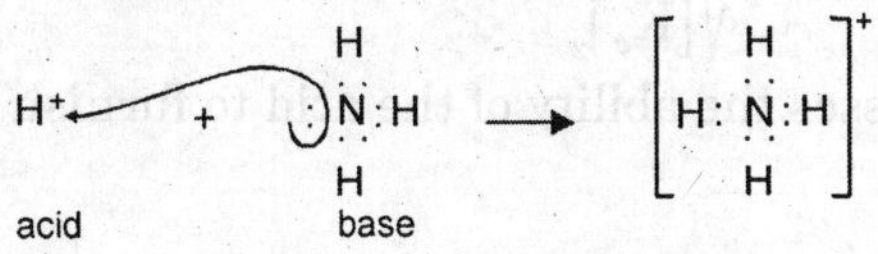

(*iii*)

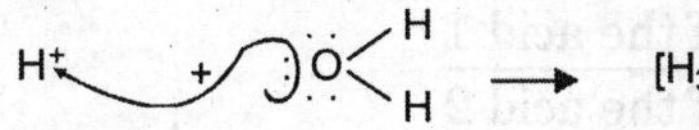

(*iv*)

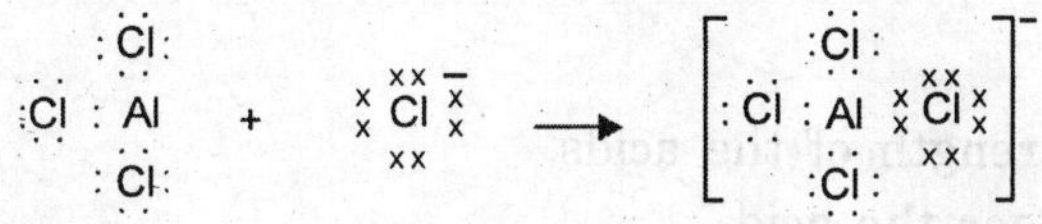

1.2.4 Dissociation Constants of Acids and Bases

(a) Acids

Ionization of an acid, HA is usually represented as,

$$HA + H_2O = H_3O^+ + A^- \qquad ...(i)$$

The equilibrium constant for this reaction is

$$K_c = \frac{[H_3O^+][A^-]}{[HA][H_2O]}$$

Concentration of water is usually large and may therefore be considered to remain essentially unchanged. Above expression may thus be written as,

$$K_c\,[H_2O] = K_a = \frac{[H_3O^+][A^-]}{[HA]} \qquad ...(ii)$$

The equilibrium constant K_a is called the **ionization** or **dissociation constant** of the acid. The ionization constant of an acid is a measure of its strength. To demonstrate this, let us examine the equilibrium (I). Suppose 'C' moles of the acids have been dissolved in one litre of water. If α is the degree of dissociation of the acid, the equilibrium concentrations of various species may be expressed as,

$$[HA] = C\,(1 - \alpha)$$
$$[H_3O^+] = C\,\alpha$$
$$[A^-] = C\,\alpha$$

According to equation (*ii*), we have

$$K_a = \frac{C\alpha \times C\alpha}{C(1-\alpha)} = \frac{C\alpha^2}{(1-\alpha)}$$

When α is small, *i.e.,* acid is weak, 1 – α = 1, so that

$$K_a = C\alpha^2$$

The equilibrium constants for two different acids at the same concentration may be expressed as,

$$[K_a]_1 = C\alpha_1^2$$

$$[K_a]_2 = C\alpha_2^2$$

α_1 and α_2 are the degrees of dissociation of the two acids considered, thus,

$$\frac{[K_a]_1}{[K_a]_2} = \frac{\alpha_1^2}{\alpha_2^2} \quad \text{or} \quad \sqrt{\frac{[K_a]_1}{[K_a]_2}} = \frac{\alpha_1}{\alpha_2}$$

The degree of dissociation of an acid expresses the ability of the acid to furnish hydrogen ions, *i.e.,* strength of the acids. Therefore,

$$\sqrt{\frac{[K_a]_1}{[K_a]_2}} = \frac{\text{Strength of the acid 1}}{\text{Strength of the acid 2}}$$

We may therefore conclude that

$$\sqrt{K_a} \propto \text{strength of the acids.}$$

Thus, larger the value of K_a stronger the acid.

(b) Bases

Ionization equilibrium involving a base B may be expressed as,

$$B + H_2O \rightleftharpoons BH^+ + OH^-$$

Assuming that the concentration of water remains unchanged, equilibrium constant K_b is written as,

$$K_b = \frac{[BH^+][OH^-]}{[B]}$$

K_b is called the ionization or dissociation constant of the base. K_b is a measure of the strength of the base larger the values of K_b stronger the base.

Following relationship exists for an acid-base conjugate pair:

$$K_a \times K_b = K_w$$

K_w is ionization constant of the solvent.

1.2.5 Ionization in Water

Water undergoes the so-called 'self ionization' as

$$H_2O + H_2O \rightleftharpoons H_3O^+ + OH^-$$

One molecule of water donates a proton *i.e.,* acts as an acid and the second acts as a base by accepting the proton. The equilibrium lies predominantly towards the left. The equilibrium constant, K_c for the above reaction may be represented by the conventional expression

$$K_c = \frac{[H_3O^+][OH^-]}{[H_2O]^2}$$

The square brackets indicate molar concentrations. The concentration $[H_2O]$ may be considered to remain essentially constant. The above equation may, therefore, be written as,

$$K_c\ [H_2O]^2 = K_w = [H_3O^+]\ [OH^-] \quad \ldots (i)$$

K_w is called the ionization constant or ionic product of water. The value of K_w at 25°C is 1×10^{-14} therefore,

$$[H_3O^+] = [OH^-] = 1 \times 10^{-7} \text{ moles/litre at 25°C.}$$

The equation (*i*) is often written as

$$K_w = [H^+]\ [OH^-]$$

K_w, which is essentially an equilibrium constant, depends on temperature.

1.3 THE pH SCALE

Hydrogen ion concentration are often preferably expressed in terms of pH. The pH of a solution is defined as "the negative logarithm of the hydrogen ion concentration" or in other words "the logarithm of the reciprocal of the hydrogen ion concentration" and is expressed in moles/litre,

$$pH = \frac{1}{\log_{10}[H^+]} = -\log_{10}[H^+] \quad \ldots (i)$$

$$[H^+] = 10^{-pH}$$

For example, if $[H^+] = 10^{-4}$

$$pH = \log\frac{1}{[H^+]}$$

then $= \log\frac{1}{10^{-4}} = \log 10^4 = 4$

In pure water, $[H^+] = 1 \times 10^{-7}$ moles/litre. The pH of water is, therefore, equal to seven. pH of 10^{-4} N HCl solution is 4 because hydrogen ion concentration is 10^{-4} moles/litre. Alkaline solutions can also be assigned pH values. We know,

$$K_w = [H^+]\ [OH^-] = 1 \times 10^{-14} \text{ at 25°C}$$

If OH^- ion concentration is known, the H^+ ion concentration can easily be calculated, since,

$$H^+ = \frac{1\times10^{-14}}{[OH^-]} \text{ moles/lt}$$

In case of 10^{-5} N NaOH solution, *e.g.*, the concentration of OH^- ions is 10^{-5} moles per litre.

Therefore,

$$H^+ = \frac{1\times10^{-14}}{10^{-5}} = 1\times10^{-9} \text{ moles/lt}$$

Thus, pH of 10^{-5} N NaOH solution is therefore equal to nine. In pure water,

$$[H^+] = [OH^-] = 1 \times 10^{-7} \text{ moles/lt.}$$

In acid solution H^+ predominate, *i.e.*,

$$[H^+] > 1 \times 10^{-7} \text{ moles/lt. or pH} < 7$$

The sign > and < stand for greater than and less than respectively.

For basic solutions,

$$[OH^-] > 1 \times 10^{-7} \text{ moles/lt.}$$

Therefore, $[H^+] < 10^{-7}$ moles/lt. Since $[H^+] \times [OH^-]$ is always equal to 1×10^{-14} at 25°C so that pH > 7. Stated otherwise, this means that a solution having the pH of which is greater than 7 will be basic in nature character. If pH of a solution is 7, the solution will be neutral.

The pH is not limited to the range 1 to 14 although such an impression is often given *e.g.*, a solution for which $[H^+]$ 1 moles/lt. will have a –ve value of pH. Such situations are, however not encountered frequently.

Expressions analogous to equation (*i*) for other quantities are also often used. Thus,

$$pOH = -\log [OH^-] \quad ...(ii)$$

$$pK_a = -\log K_a \quad ...(iii)$$

$$pK_b = -\log K_b \quad ...(iv)$$

$$pK_w = -\log K_w \quad ...(v)$$

It can easily be shown that sum of pH and pOH in water or aqueous solution is always equal to 14 at 25°C. We know

$$K_w = [H^+][OH^-] \quad ...(vi)$$

$$\therefore \quad -\log K_w = -\log [H^+] - \log [OH^-] \quad ...(vii)$$

Using equations (*v*), (*i*) and (*ii*), we have

$$pK_w = pH = pOH$$

Since $$K_w = 1 \times 10^{-14} \text{ at } 25°C \quad ...(viii)$$

$$pK_w = 14$$

Thus, it follows that,

$$\therefore \quad pH + pOH = 14$$

Calculation of pH

In the dissociation of the weak acid HA,

$$HA \rightleftharpoons H^+ + A^-$$

the concentrations of H^+ and A^- ions are equal. Therefore, we may write

$$K_a = \frac{[H^+][A^-]}{[HA]}$$

or $$[H^+][A^-] = K_a [HA]$$

Since $$[H^+] = [A^-]$$

we have $$[H^+]^2 = K_a [HA]$$

or $$[H^+] = \sqrt{K_a [HA]}$$

Example 1: Calculate the pH of 0.1 N acetic acid. The K_a for acetic acid is 1.86×10^{-5}.

Solution: Acetic acid is a week acid. CH_3COOH may therefore be taken to be equal to 0.1 gm moles/lt.

$$\text{Now,} \quad H^+ = \sqrt{K_a.[CH_3COOH]} = \sqrt{1.86\times10^{-5}\times0.1}$$

$$= \sqrt{1.86\times10^{-6}}$$

$$\log[H^+] = \log\sqrt{1.86\times10^{-6}} = \frac{1}{2}[\log 1.86 + \log 10^{-6}]$$

$$= \frac{1}{2}[0.2695-6]$$

$$= -2.86$$

$$\therefore \quad pH = 2.86$$

Example 2: Calculate the pH value of a solution whose H^+ ion concentration is 0.005 gm equivalent per litre.

Solution:

$$pH = -\log [H^+]$$

$$= -\log [0.005]$$

$$= -(-3.699)$$

$$= -(-2.301)$$

$$= 2.301$$

Example 3: Calculae the pH values of the following solutions assuming complete ionization.

(*a*) 4.8×10^{-4} N Acid

(*b*) 0.0016 N Base

Solution:

(*a*) $[H^+] = 4.9 \times 10^{-4}$ moles/lt.

$$\therefore \quad pH = \log\frac{1}{[H^+]} = \log\frac{1}{4.9\times10^{-4}}$$

$$= \log\frac{10^4}{4.9}$$

$$pH = \log 10^4 - \log 4.9 = 4 - 0.69$$

$$= 3.31$$

(*b*)

$$[H^+] = \frac{10^{-14}}{[OH^-]} = \frac{10^{-14}}{1.6\times10^{-3}} = \frac{10^{-11}}{1.6} \text{ moles/lt.}$$

$$\text{Now} \quad pH = \log\frac{1}{[H^+]} = \log\frac{1.6}{10^{-11}}$$

$$= \log 1.6 + \log 11$$

$$= 0.20 + 11$$

$$= 11.20$$

1.4 BUFFER SOLUTIONS

pH of an aqueous solution changes whenever an acid or alkali is added *e.g.*, if we add a small amount of an acid (HCl) to an aqueous solution, its pH decreases due to increase in concentration of hydrogen ions. On the other hand, if a small amount of a base (NaOH) is added, the pH of the solution increases. There are however, some solutions which resists alterations in pH when a small amount of a strong acid or alkali is added.

It is noticed, however, that a solution containing a weak acid and the salt of it with a strong base (*e.g.*, $CH_3COOH + CH_3COONa$) or a weak base and the salt of it with a strong acid (*e.g.*, $NH_4OH + NH_4Cl$) possesses the following characteristics:

(*i*) It has a definite pH values.

(*ii*) Its pH value does not alter either on keeping for long or on dilution.

(*iii*) Its pH value is very slightly altered by the addition of a strong acid in the first case, *viz.* $CH_3COOH + CH_3COONa$ and a strong base in the second case *viz*,

$$NH_4Cl + NH_4OH.$$

Such solutions with reserve acidity or alkalinity which resist a change in pH on the addition of a small amount of acid or base are called buffer solutions.

Most of the buffer solutions are mixtures of weak acids and their salts or weak bases and their salts.

Buffer Solutions of a Weak Acid and its Salts: We consider a typical buffer solution containing a weak acid, CH_3COOH and its salt CH_3COONa since acetic acid is a weak acid, there will be very few H^+ ions in solution but due to the presence of the strong electrolyte, CH_3COONa, we have an excess of Na^+ and CH_3COO^- ions. When a strong acid (HCl) is added, the H^+ ions of it combine with the CH_3COO^- ions to form the feebly ionised CH_3COOH with the result that the pH value does not alter to any considerable extent.

$$CH_3COO^- + H^+ \rightleftharpoons CH_3COOH$$

In case, however, when a strong base is added, the OH^- ions furnished by the base are neutralised by the CH_3COOH present and there is little or no change in the pH value of the solution.

$$CH_3COOH + OH^- \rightarrow CH_3COO^- + H_2O$$

Buffer Solutions of Weak Base and its Salts: Consider a buffer solution consisting of a mixture of the weak base (NH_4OH) and its salt NH_4Cl such a solution contains a large concentration of unionised molecules of NH_4OH along with NH_4^+ ions furnished by NH_4Cl.

When a strong acid is added to such a solution, the H^+ ions thus introduced are neutralised by NH_4OH

$$H^+ + NH_4OH \rightarrow NH_4^+ + H_2O$$

and there is no change in pH value. On the addition of a bases, the OH^- ions furnished by it are fixed up by NH_4^+ ions to form the feebly ionised NH_4OH and there is practically no change in the pH value of the solution.

$$OH^- + NH_4^+ \rightarrow NH_4OH$$

pH value of Buffer Solution (Preparation of Buffer Solution): The dissociation of a weak acid HA according to the equation

$$HA \rightarrow H^+ + A^-$$

is governed by the expression.

$$K_a = \frac{[H^+][A^-]}{[HA]}$$

$$\therefore \qquad [H^+] = K_a \frac{[HA]}{[A^-]}$$

for preparing a buffer solution of a given pH, it is first necessary to select a weak acid [HA] whose pK_a value is as near as possible to the required pH so as to obtain a maximum buffer capacity. A salt of this weak acid with a strong base (NaA) is then added to it. Let C_1 and C_2 be the molar concentrations of the acid and the salt respectively. The feeble dissociation of the weak acid is further suppressed by the presence of A^- ions with the result that the concentration of the unionised acid may be taken as the same as the concentration of the total acid *i.e.,* $[HA] = C_1$.

The concentration of $[A^-]$ ions may likewise be presumed to be equal to the concentration of the salt it is entirely dissociated, *i.e.,* $[A^{-1}] = C_2$.

Substituting these values in the above equation, we have

$$[H^+] = K_a \frac{C_1}{C_2}$$

$$= K_a \frac{[Acid]}{[Salt]}$$

Taking logarithms and reversing the sign

$$-\log [H^+] = -\log K_a + \log \frac{[Salt]}{[Acid]}$$

or

$$pH = pK_a + \log \frac{[Salt]}{[Acid]}$$

Likewise for a mixture of a weak base and its salt it can be shown that

$$pOH = pK_b + \log \frac{[Salt]}{[Base]}$$

This is known as Henderson's Equation.

Example 4: Calculate the pH of a buffer solution containing 0.1 mole of CH_3COOH and 0.01 mole of CH_3COOK per litre. The dissociation constant of CH_3COOH is 1.84×10^{-5}.

Solution:

$$pH = pK_a + \log \frac{[Salt]}{[Acid]}$$

$$pK_a = -\log K_a = -\log (1.84 \times 10^{-5})$$

$$= 4.735$$

$$[Salt] = 0.01 \text{ mole/lt.}$$

$$[Acid] = 0.1 \text{ mole/lt.}$$

$$\therefore \quad pH = 4.735 + \log_{10} \frac{0.01}{0.1}$$
$$= 4.735 + \log_{10} 10^{-1}$$
$$= 4.735 - 1$$
$$= 3.735$$

1.5 SOLUBILITY PRODUCT

When a slightly soluble ionic solid such as AgCl is placed in water, Ag^+ and Cl^- from solid phase pass into solution till the solution becomes saturated. Now there exists an equilibrium between the ions present in the saturated solution and the ions present in the solid phase. Thus,

$$AgCl\,(s) \rightleftharpoons Ag^+ + Cl^-$$

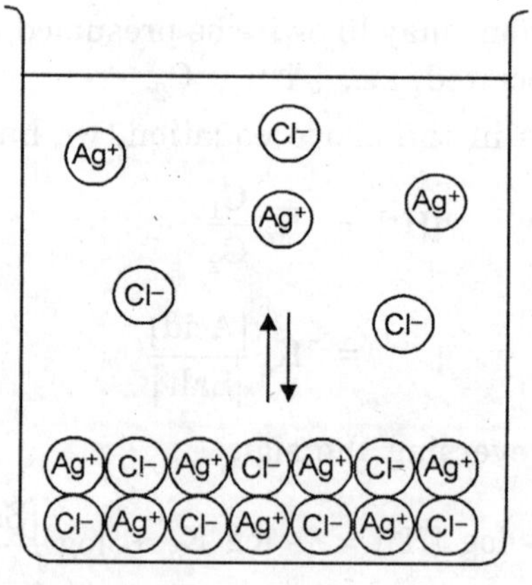

Fig. 1.2: Equilibrium between an ionic solid and saturated solution

Applying the law of mass action,

$$\frac{[Ag^+][Cl^-]}{[AgCl_{solid}]} = K, \text{ at constant temperatures}$$

Since the concentration of solid AgCl in the solid phase (moles of AgCl per litre of solid phase) is fixed, no matter how much solid is present in contact with solution, we can write,

$$[Ag^+]\,[Cl^-] \rightleftharpoons [AgCl_{solid}] \times K = K_{sp}.$$

K_{sp} is the **solubility product constant** and the expression $[Ag^+]\,[Cl^-]$ the ionic product. In other words, the product of the concentration of ions in the saturated solution of a spairingly soluble salt as AgCl is constant at a given temperature. This principle is known as the Solubility Product Principle.

Example 5: You are given a solution of AgBr in which the concentration of Ag^+ ion is 1×10^{-6} moles/lt. Find the minimum concentration of Br^- ions necessary to cause precipitation of AgBr. The solubility product for AgBr = 4×10^{-13}.

Solution :

$$[Ag^+]\,[Br^-] = K_{sp}$$
$$= 4 \times 10^{-13}$$

Here $[Ag^+] = 1 \times 10^{-6}$ moles/litre

Thus, $(1 \times 10^{-6}) \times [Br^-] = 4 \times 10^{-13}$

$$[Br^-] = \frac{4 \times 10^{-13}}{1 \times 10^{-6}}$$

$$= 4 \times 10^{-7} \text{ moles/litre.}$$

1.6 CONDUCTORS

Substances are divided into two classes, *viz.*, conductors and non-conductors or insulators. Conductor is a substance which allows electric current to pass through it, *e.g.*, metals, sulphides such as pyrites, galena, aquous solution of acids, bases and salts, fused salts and impure water. Conductors are of two types:

(*i*) Metallic conductors

(*ii*) Electrolytic conductors or electrolytes.

1.6.1 Metallic Conductors

Metallic conductors are those in which electricity flows without producing chemical changes. In this case conduction takes place by the movement of electrons under an applied potential. Metals and graphites are the examples of metallic conductors. The passage of an electric current through a copper wire is an example of metallic conduction. In metallic conduction:

- electrons flow from –ve to +ve end
- no change in the composition of conductor takes place
- no transfer of matter takes place.
- resistance to the flow of current increases with the increase in temperature and hence increase in temperature decreases the conduction.

1.6.2 Electrolytic Conductors

Those substances whose water solution conducts the electric current and which are decomposed by the passage of electric current are called electrolytic conductors or electrolytes.

In this case conduction takes place by the movement of ions. Solution of salts, acids and bases are the examples of electrolyte. Electrolytes also conduct electricity in fused state and undergo decomposition by the passage of electric current. Substances whose aquous solution does not conduct electricity are called non-electrolytes. They do not conduct electricity in the fused state. Solutions of cane sugar, alcohol, glycerine, glucose, urea, *etc.*, are the examples of non-electrolytes.

In electrolytic conduction:

- ions move towards oppositely charged electrodes
- chemical changes takes place at the electrodes
- transfer of matter takes place in the form of ions
- the resistance to the flow of current decreases with increase in temperature and hence increase in temperature increases the conduction

PROBLEMS

1. Write the short notes:
   (a) Bronsted Concept of Acids and Bases.
   (b) Buffer solutions.
   (c) pH value.
2. Calculate the pH of a solution consisting 0.05 molal lactic acid and 0.122 molal Na lactate.
   (K_a = 1.37 × 10^{-4})
3. Calculate pH of solution consisting 0.1 M of NH_4OH and 0.24 M NH_4Cl.
   (K_b = 1.8 × 10^{-5})
4. 10 cc of 0.1 M NaOH solution added to 50 cc of 1.1 M acetic acid solution. What will be the pH of the resulting solution?
   (K_{HAC} = 1.75 × 10^{-5})
5. A solution which is 0.05 M in respect of benzoic acid and 0.1 M in sodium benzoate, has pH = 4.50. Find dissociation constant of the acid.
6. pK_a for acetic acid is 4.76. How will you prepare a buffer solution of pH 4.40 by mixing sodium acetate and acetic acid?
7. Explain the Lewis concept of acids and bases. Identify acids and bases in the following: NH_3, H_3O^+, BF_3, $ZnCl_2$, $C_2H_5NH_2$, SO_3, H_2O, ROH.
8. Establish the relation between pH and pOH. What is the pH of solution containing the H^+ concentration of 3 × 10^{-7} moles per litre?
9. What are conjugate acid and base? Give the conjugate base of the following:
   NH_4^+, HCl, H_3PO_4, H_2S, HCO_3^-.

CHAPTER 2

CATALYSIS

2.1 INTRODUCTION

A substance which increases the rate of a chemical reaction without being consumed in the process and remains unchanged chemically and quantitatively is called a *Catalyst* and the phenomenon is known as *Catalysis*. A catalyst can either accelerate or retard the reaction.

The term catalysis is used generally to express acceleration. Therefore a catalyst is defined as a substance that changes the reaction rates without affecting the overall energetics of the reaction. *e.g.,*

(*i*) Hydrogen peroxide normally decomposes slowly into oxygen and water

$$2H_2O_2 \xrightarrow[\text{Catalyst}]{\text{Pt}} 2H_2O + O_2$$

But in the presence of even a trace amount of finely powdered platinum, the decomposition becomes explosive.

(*ii*) Decomposition of potassium chlorate ($KClO_3$) is enhanced by the addition of a relatively small quantity of manganese dioxide (MnO_2)

$$2KClO_3 \xrightarrow[\text{Catalyst}]{MnO_2} 2KCl + 3O_2$$

Platinum and manganese dioxide used in the above reactions are catalysts.

Catalysts do not appear in the product of the reaction and may also be recovered at the end of the reaction. In a reversible reaction, the catalyst increases the rate of both forward and backward reactions to the same extent and therefore does not alter the position of chemical equilibrium.

2.2 CHARACTERISTICS OF CATALYSTS

Following are the main characteristics of a catalyst:

(*i*) The catalyst is unchanged chemically and quantitatively at the end of the reaction.

(*ii*) The catalyst does not alter the position of equilibrium in a reversible reaction.

(*iii*) The catalyst does not initiate a reaction that would not otherwise occur.

(*iv*) Only a little amount of catalyst is sufficient to bring about a considerable extent of reaction.

(*v*) The temperature effects on the reaction rates remains unaltered even in the presence of a catalyst.

(*vi*) The activity of catalytic substances can be changed by the addition of even a trace amount of some foreign substances, known as catalytic promoter and catalytic poisons.

(*vii*) The catalysts are very specific in nature to chemical reactions. Therefore, the mechanism and the nature of the products depend to a large extent on the nature of the catalyst *e.g.,*

Ethanol vapour at 350°C, dehydrogenated in the presence of nickle catalyst while in the presence of alumina they undergo dehydration.

$$C_2H_5OH \xrightarrow[350°C]{Ni} CH_3CHO + H_2 \text{ (Dehydrogenation)}$$

$$C_2H_5OH \xrightarrow{Al_2O_3} C_2H_4 + H_2O \text{ (Dehydration)}$$

2.3 TYPES OF CATALYSIS

There are two types of catalysis:

(*i*) Homogeneous catalysis and
(*ii*) Heterogeneous catalysis

2.3.1 Homogeneous Catalysis

All such catalytic reactions where in all reactants, products and the catalysts itself are in the same physical phase, are known as homogeneous catalysis *e.g.,*

(*i*) Liquid phase homogeneous catalysis is the inversion of sucrose dissolved in water. This reaction which yields a mixture of glucose and fructose is catalysed by a drop of a strong acid such as H_2SO_4:

$$\underset{\text{Sucrose}}{C_{12}H_{22}O_{11}} + \underset{\text{Water}}{H_2O} \xrightarrow[\text{Catalyst}]{H_2SO_4} \underset{\text{Glucose}}{C_6H_{12}O_6} + \underset{\text{Fructose}}{C_6H_{12}O_6}$$

In the above reaction the reactants; sucrose and water, products; glucose and fructose and catalyst; sulphuric acid are in the same physical phase *i.e.,* aqueous phase.

(*ii*) In the Lead Chamber process for the manufacture of sulphuric acid, the oxidation of sulphur dioxide (SO_2) to sulphur trioxide (SO_3) is catalysed by nitric oxide (NO_2).

$$2SO_2 + O_2 \xrightarrow{NO_2} 2SO_3$$

In the above reaction the reactants (SO_2 and O_2), product (SO_3) and the catalyst (NO_2) are in the same physical phase *i.e.,* gaseous phase.

Some more examples of homogeneous catalysis are the reactions catalysed by acids or bases and thus are classified as acid-base catalysis. Some such examples are given below:

2.3.1.1 Acid Catalysis

The hydrolysis of esters, amides, diazo acetic ester and iodination of acetone are catalysed by acid *e.g.,*

(i) $$\underset{\text{Ethyl acetate}}{CH_3COOC_2H_5} + H_2O \xrightarrow[\text{Cat.}]{H^+} \underset{\text{Acetic acid}}{CH_3COOH} + \underset{\text{Ethanol}}{C_2H_5OH}$$

(ii) $$\underset{\text{Acetamide}}{CH_3CONH_2} + H_2O \xrightarrow[\text{Cat.}]{H^+} \underset{\text{Acetic acid}}{CH_3COOH} + \underset{\text{Ammonia}}{NH_3}$$

(iii) $$\underset{\text{Acetone}}{CH_3COCH_3} + I_2 \xrightarrow[\text{Cat.}]{H^+} CH_3COCH_2I + HI$$

2.3.1.2 Base Catalysis

The dissociation of diacetone alcohol, decomposition of nitramide and depolymerisation of diacetone alcohol are well known examples of base catalysed reactions. *e.g.*,

(i) $$\underset{\text{Diacetone alcohol}}{CH_3C(OH)CH_3CH_2COCH_3} \xrightarrow[\text{Cat.}]{OH^-} \underset{\text{Acetone}}{2CH_3\,COCH_3}$$

(ii) $$\underset{\text{Nitramide}}{NH_2NO_2} \xrightarrow[\text{Cat.}]{OH^-} NO_2 + H_2O$$

The process of acid-base catalysis has led to the development of the proton transfer concept of acids and bases.

It is of interest to note that the role of the catalysts in these reactions is the lowering of the activation energy by providing an alternative path for the reaction. The rate of these reactions are always found to be dependent on the catalyst concentration.

2.3.2 Heterogeneous Catalysis

All such processes in which the catalysts, reactants and the products are not in the same physical phase are known as heterogeneous catalysis. Some well known examples of heterogeneous catalysis are given as follows:

(i) Manufacture of sulphuric acid (contact process)

The essential part of this process is the oxidation of SO_2 to SO_3

$$SO_2 + \frac{1}{2}O_2 \xrightarrow[\text{Cat.}]{Pt} SO_3$$

In the above reaction, the reactants, SO_2 and O_2 and the product, SO_3 are in the gaseous phase whereas the catalyst, platinum is in solid phase.

(ii) Manufacture of ammonia (Haber process)

$$N_2 + 3H_2 \xrightarrow[\text{Cat.}]{Fe_2O_3} 2NH_3$$

In the above reaction the reactants, N_2, H_2 and the product, NH_3 are in gaseous phase while the catalyst Fe_2O_3 is in solid phase.

(iii) Manufacture of Hydrogen

$$CH_4 + H_2O \xrightarrow[750-900°C]{Ni} CO + 3H_2$$

$$CO + H_2O \underset{400-450°C}{\overset{Fe}{\rightleftharpoons}} CO_2 + H_2$$

(iv) Manufacture of methanol and ethanol

$$CO + 2H_2 \xrightarrow[400°C]{\text{Zn/Cr Oxides}} CH_3OH$$

$$CH_2 = CH_2 + H_2O \xrightarrow[300°C \ \& \ 7 \times 10^6 pa]{\text{Phosphoric acid}} C_2H_5OH$$

(*v*) The cracking of petroleum

Petroleum, which consists mainly of saturated paraffin hydrocarbons, is broken down into smaller molecules under the influence of heat and a suitable catalyst, *i.e.*, mixture of alumina and silica.

2.4 THEORIES OF CATALYSIS

There are two main theories of catalytic reactions:

(*i*) Intermediate Compound formation theory and

(*ii*) Adsorption theory

2.4.1 Intermediate Compound Formation Theory

The accelerating influence of catalysts in heterogeneous catalysis is usually explained by the formation of *Intermediate Compounds.* According to this theory, the reaction is brought about by a path involving an intermediate formed from the catalyst and one of the reactants. The intermediate compound is highly reactive and reacts with the other reactants or is unstable and decomposes into the products with the regeneration of the catalyst.

Let us consider a general case wherein reactants *X* and *Y* form product *XY* very slowly

$$X + Y \rightleftharpoons XY$$

But in the presence of catalyst K, the reaction occurs rapidly as follows:

$$X + K \longrightarrow XK \quad ...(i)$$

$$XK + Y \longrightarrow XY + K \quad ...(ii)$$

$$X + Y + K \longrightarrow XY + K$$

The above scheme does not result in any change in the catalyst. It is a fact that catalytic action is proportional to the amount of catalyst present. *e.g.*,

(*i*) Manufacture of sulphuric acid (lead chamber process)

$$2SO_2 + O_2 + 2H_2O \longrightarrow 2H_2SO_4$$

Normally process proceeds very slowly. However, in the presence of nitric oxide as a catalyst, the reaction occurs rapidly.

$$2NO + O_2 \longrightarrow 2NO_2 \quad ...(i)$$

$$2NO_2 + 2SO_2 + 2H_2O \longrightarrow 2H_2SO_4 + 2NO \quad ...(ii)$$

$$2SO_2 + O_2 + 2H_2O + 2NO \longrightarrow 2H_2SO_4 + 2NO$$

In this reaction actually an intermediate compound Nitroso-Sulphuric acid is formed.

(*ii*) Manufacture of diethyl ether from ehtyl alcohol in the presence of catalyst H_2SO_4.

$$C_2H_5OH + H_2SO_4 \longrightarrow \boxed{C_2H_5HSO_4} + H_2O$$

$$C_2H_5HSO_4 + C_2H_5OH \longrightarrow C_2H_5OC_2H_5 + H_2SO_4$$

$$2\ C_2H_5OH + H_2SO_4 \longrightarrow C_2H_5OC_2H_5 + H_2O + H_2SO_4$$

The intermediate compound ethyl hydrogen sulphate is first formed which subsequently reacts with a molecule of ethyl alcohol to form ether and regenerates sulphuric acid.

2.4.2 Adsorption Theory

When the surface of a solid is exposed to a gas, the gas molecules tend to adhere to the surface. It is now known that all solids in contact with a gas accumulate a film of gas molecules on the surface. The forces that cause the molecules to adhere to the surface may be either the normal force of attraction between molecules *i.e.*, Van der Waals Forces (VWF) or chemical bond formed between the molecules and the surface. This phenomenon is known as *Adsorption*. The substance which adheres to the solid is called **'adsorbate'** while the solid is called **'adsorbent'**.

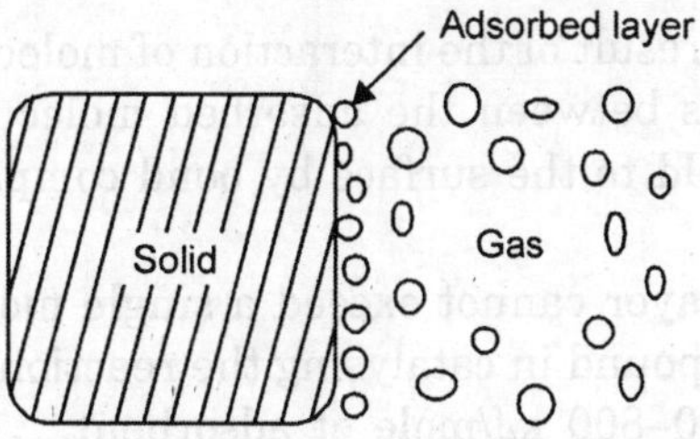

Fig. 2.1: Adsorption of gas molecules on solid surface

In most of the reactions involving heterogeneous catalysis, action of a solid catalyst is explained on the basis of the phenomenon of adsorption.

The adsorption of molecules at a solid surface is of the two types:

(*i*) Physical adsorption or physisorption

(*ii*) Chemical adsorption or chemisorption

2.4.2.1 Physical Adsorption

When an atom or a molecule is bound to the surface by Van der Waals forces *i.e.*, intermolecular attraction, the phenomenon is called as physical adsorption. The physically adsorbed layer of molecules can easily be removed. The heat of adsorption is of the order of 8–20 kJ/mole of the adsorbate.

2.4.2.2 Chemical Adsorption

Adsorption by chemical forces is known as chemical adsorption and here there is a chemical bond formed between the adsorbed molecule and the surface.

At the surface of a solid catalyst, there exists free valencies, as shown below:

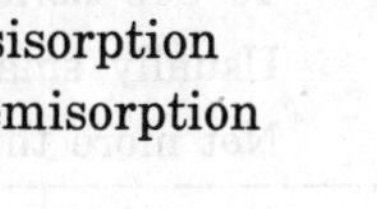

Fig. 2.2: Free valencies at the solid surface

The ways in which molecules can be adsorbed on the surface of a metal are shown in the Fig. 2.3 given below:

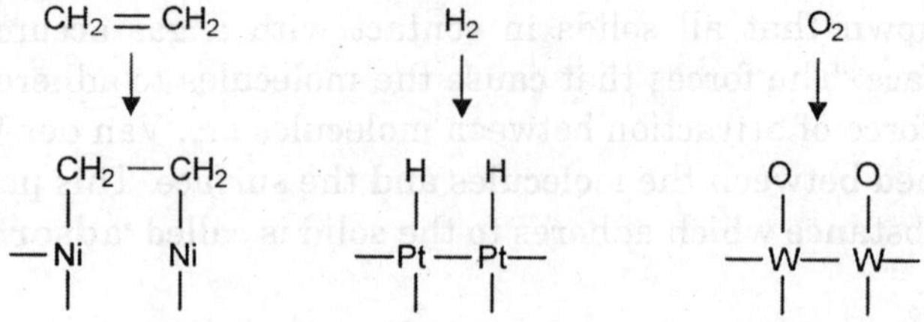

Fig. 2.3: Chemical adsorption of molecules on metal

Chemisorption occurs as a result of the interaction of molecules with these free valencies. It involves sharing of electrons between the adsorbed molecules and the catalyst surface. The adsorbed molecules are held to the surface by bond comparable with those of chemical compounds.

The chemically adsorbed layer cannot exceed a single molecule in thickness and plays the role of an intermediate compound in catalysing the reaction. The energy of chemisorption is commonly in the range of 40–800 kJ/mole of adsorbent.

Chemical adsorption and physical adsorption can be distinguished on the basis of following criteria:

Sl. No.	*Criteria*	*Chemisorption*	*Physisorption*
1.	Heat of adsorption	40–800 kJ/mole	8-20 kJ/mole
2.	Activation of energy	Usually small	Zero
3.	No. of layers adsorbed	Not more than one	More than one

The mechanism of surface catalysis is explained in terms of diffusion theory of catalysis, which postulates the following sequence of events for a gaseous reaction on solid surface.

(*i*) Diffusion of the reactants to the surface
(*ii*) Adsorption of the reactant molecules on to the surface
(*iii*) Actual chemical reaction on the surface
(*iv*) Desorption of the products from the surface
(*v*) Diffusion of the products away from the surface.

2.5 ACTIVATION ENERGY AND CATALYSIS

On the basis of thermodynamic considerations, we know that for all chemical reactions that occur, the final state of the system must have a lower free energy than the initial state. The rate at which a chemical reaction occurs depends mainly on the activation energy. The minimum amount of energy required to cause a chemical reaction is known as the activation energy. A catalyst can either increase or lower the activation energy of the reaction by providing a new path.

A reaction occurs as a result of collisions between the reacting molecules. The collisions are not very effective at normal temperature, because the molecules do not possess enough energy. When temperature of the system increases, the kinetic energy of the molecules increases, but the molecules do not reach the energy barrier unless they attain the minimum amount of energy called energy of activation. The activated molecules first form an activated complex on collision and thus activated complex breaks to yield products.

When a catalyst is present, it normally decreases the energy of activation of the reaction by providing a new path.

A positive catalyst speeds up the reaction by lowering the activation energy whereas a negative catalyst slows down the reaction by increasing the activation energy.

The schematic representation of the effect of a catalyst with the help of a potential energy diagram is shown below in Fig. 2.4.

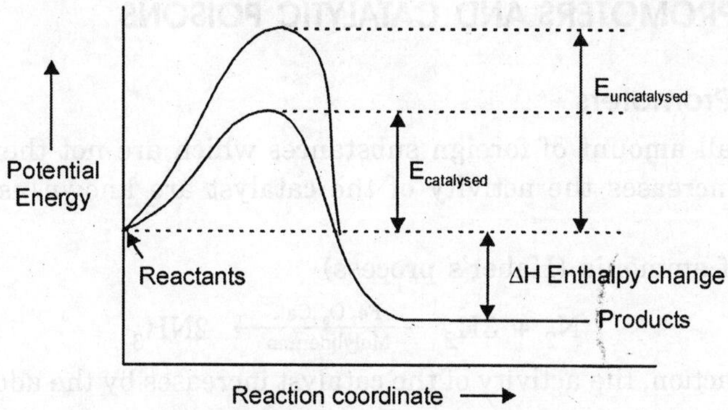

Fig. 2.4 Potential energy diagram

It is clear from the above diagram that a catalyst provides an alternative path of lower energy barrier. The enthalpy change (ΔH) however is the same for both catalysed and uncatalysed reaction. The activation energy for catalysed and uncatalysed reactions are given below:

Sl. No.	*Reaction*	*Activation Energy*	
		Uncatalysed Reaction	Catalysed Reaction
1.	Pyrolysis of $C_2H_5OC_2H_5$	53.50 kJ/mole	34.30 kJ/mole
2.	Prolysis of CH_3CHO	45.50 kJ/mole	32.50 kJ/mole

It is clear from the above that a catalyst acts by lowering the energy of activation.

2.6 POSITIVE AND NEGATIVE CATALYSTS

2.6.1 Positive Catalyst

When a catalyst accelerates the speed of chemical reaction it is known as a positive catalyst. *e.g.*,

(*i*) Decomposition of potassium chlorate

$$2KClO_3 \xrightarrow[\text{Cat.}]{MnO_2} 2KCl + 3O_2$$

decomposition of $KClO_3$ is enhanced by the addition of a catalyst such as MnO_2.

(*ii*) Manufacture of sulphuric acid

$$2SO_2 + O_2 \xrightarrow[\text{Cat.}]{Pt} 2SO_3$$

In the above reaction rate of reaction accelerates by the addition of Pt as a catalyst. MnO_2 and Pt used in the above reactions are called as positive catalyst.

2.6.2 Negative Catalyst

When a catalyst retards the speed of a chemical reaction it is known as a negative catalyst. *e.g.*,

Alcohol retards the oxidation of chloroform to poisonous phosgene.

2.7 CATALYTIC PROMOTERS AND CATALYTIC POISONS

2.7.1 Catalytic Promoters

The addition of small amount of foreign substances which are not themselves catalytically active, sometimes increases the activity of the catalyst are known as catalytic promoters *e.g.*,

Manufacture of ammonia (Haber's process).

$$N_2 + 3H_2 \xrightarrow[\text{Molybdenum}]{Fe_2O_3\ \text{Cat.}} 2NH_3$$

In the above reaction, the activity of the catalyst increases by the addition of molybdenum which acts as catalytic promoters.

2.7.2 Catalytic Poisons or Anti-Catalysts

The activity of a catalyst is inhibited or completely destroyed by the presence of even minute traces of certain substances, such substances are known as catalytic poisons. *e.g.*,

Manufacture of sulphuric acid (contact process)

$$SO_2 + O_2 \xrightarrow[As_2O_3]{Pt} SO_3$$

In the above reaction addition of small quantity of As_2O_3 inhibited the activity of the catalyst (Pt), hence called a catalytic poison.

2.8 AUTOCATALYSIS

When a product formed in the course of reaction, (sometimes one of the reaction reactants) enhances the velocity of the reaction and acts as a catalyst is known as autocatalyst and the phenomenon is known as autocatalysis. *e.g.*,

Hydrolysis of an ester

$$\underset{\text{Ester}}{RCOOR'} + H_2O \rightleftharpoons \underset{\text{Acid}}{RCOOH} + \underset{\text{Alcohol}}{R'OH}$$

Since the acid (RCOOH) formed as a result of hydrolysis catalyses the reaction.

2.9 ENZYME CATALYSIS

The important biochemical characteristics of a living cell is its ability to achieve a large number of rapid chemical conversions. This is achieved by substances known as enzymes or ferments. Enzymes are complex protein substances produced by living cells. They can catalyse numerous chemical reactions occurring in animal or plant bodies. They split molecules, join them or transfer group or atoms from one molecule to another.

On the basis of catalytic functions, enzymes have been classified into six major groups.

Sl. No.	*Catalytic Function*	*Enzyme*
1.	Oxidation reduction	Oxido-reductase
2.	Group transfer	Transferases
3.	Hydrolysis	Hydrolases
4.	Group removal	Lyases
5.	Isomerisation	Isomerases
6.	Joining of molecules	Ligases

Biological reactions are often catalysed by enzymes only in the presence of an extra non-protein molecule. These are called coenzymes. Some important coenzymes are given below:

Sl. No.	*Enzymes*	*Coenzymes*	*Non-Protein Molecules*
1.	Flavin adenin dinucleotide (FAD)	Cytochromes	Riboflavin
2.	Nicotinamide adenine dinuclpyotide (NAD)	Dehydrogenases	Nicotinamide
3.	Pyridoxal phosphate	Trans-amination	Pyridoxin
4.	Thiamine Pyrophosphate	Dicarboxylases	Thiamine

Enzymes increase the velocity of chemical reactions by reducing the energy of activation like other catalysts, *e.g.*,

Reaction	*Catalyst*	*E(kJ/mole) Activation Energy*	*Reaction Velocity*
Decomposition	(*i*) None	18.00	1
of H_2O_2	(*ii*) Platinum	12.00	26.00
	(*iii*) Catalase	0.60	above 10^{10}
Hydrolysis	(*i*) Acid	26.00	1
of Sucrose	(*ii*) Invertase	11.00	above 10^{11}

Now it is clear from the table that the enzyme catalase and invertase are more efficient catalysts compared to inorganic catalysts like platinum and acid because they cause greatest reduction in the activation energy. The lowering of activation energy in such cases are not very large. However, there is a tremendous increase in the velocity of the reactions.

In an enzymatically catalyzed reaction, the enzyme combines with the substrate and forms enzyme substrate complex which subsequently decomposes into products and the free enzyme. In this process, three steps occur successively at the surface. These are:

(*a*) A molecule is adsorbed at the surface and is attached to an active site.

(*b*) It decomposes while on the site.

(c) Products are desorbed from the surface which gives the site further opportunity to adsorb fresh reactants.

The enzymatically catalyzed reaction can be expressed by the following schemes:

$$\underset{\text{(Enzyme)}}{E} + \underset{\text{(Substrate)}}{S} \rightleftharpoons \underset{\text{(Enzyme-substate complex)}}{ES}$$

$$ES \rightarrow \underset{\substack{\text{Free}\\\text{(Enzyme)}}}{E} + \underset{\text{(Product)}}{P}$$

2.9.1 Characteristics of Enzyme Catalysis

Following are the main characteristics of the enzyme catalysis:

(i) The enzymes are not destroyed and are regenerated at the end of the reaction.

(ii) The enzymes are inhibited or inactivated at high temperature, extreme of pH, UV rays, heavy metal ions and mechanical agitation.

(iii) Enzyme reactions are reversible.

(iv) Enzyme reactions are much sensitive to catalytic poisons such as HCN, H_2S, CS_2 etc.

(v) Enzymes are highly specific and each enzyme catalyses a particular reaction. *e.g.*, an enzyme invertase can break sucrose into glucose and fructose but fails to break up a similar disaccharide *i.e.*, maltose, which requires another enzyme *i.e.*, maltase.

(vi) In certain cases the activity of an enzyme depends upon certain non-protein substances called co-enzymes. For each enzyme there is only one co-enzyme.

2.9.2 Factors Affecting Enzyme Activity

Following are the main factors which affect the enzyme activity:

(i) Substrate concentration

(ii) Temperature

(iii) pH

(iv) Oxidation-reduction potential

(v) Inhibitors

2.9.2.1 *Substrate Concentration*

Upto a certain concentration, the rate of formation of end products increases with increasing concentration of the substrate. But the point where substrate is saturated with the enzyme, further increase in substrate concentration has no influence on the rate of formation of end products.

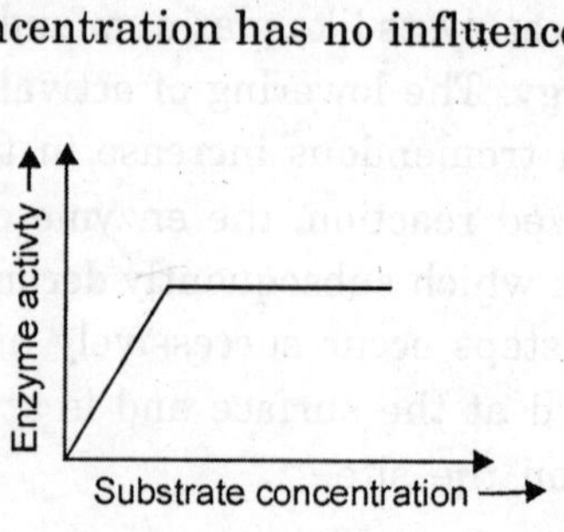

2.9.2.2 *Temperature*

The above figure shows the general effect of temperature on the enzyme catalysed reactions. In general the enzymes show a temperature optimum which is usually between 30° to 50°C.

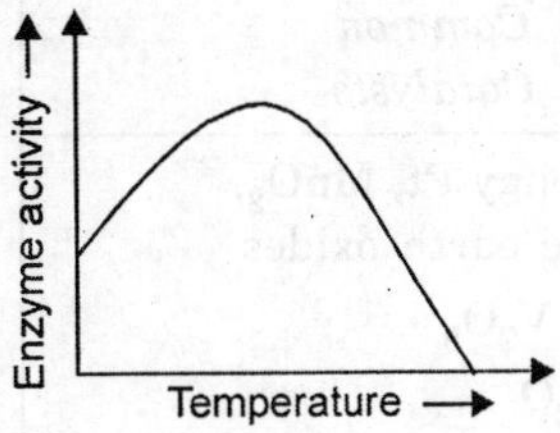

2.9.2.3 *pH*

The enzymes are sensitive to change in pH of the medium. Maximum activity is usually observed near the iso-electric point (IEP), *e.g.*, trypsin, which has its IEP between pH 7 and 8 shows greatest activity towards its substrates in the pH range of 7 to 9. The activity of this enzyme rapidly drops to zero when the pH is raised or lowered by about 2 pH units from the pH optimum.

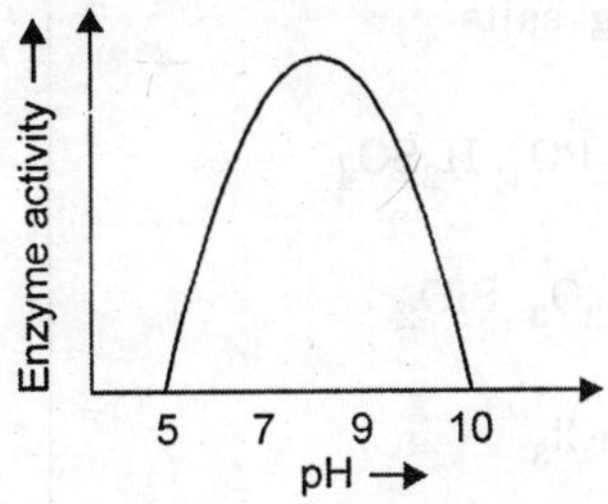

2.9.2.4 *Oxidatjon-Reduction Potential*

Many enzymes are sensitive to the action of oxidising or reducing agent. Changes in the redox potential occurs because of the presence of –SH group in the enzyme molecule.

2.9.2.5 *Inhibitors*

There are many substances which inhibit the enzyme activity are called inhibitors.

2.10 INDUSTRIAL CATALYSTS

Catalysts find extensive use in the chemical industry, and are used in gaseous, liquid or solid forms. A few examples of catalyst that are of major importance in chemical industry are Ni, Fe and Pt are used to accelerate reactions in which hydrogen is added to another atom or compound. The synthesis of ammonia and the conversion of cotton seed oil to yield edible fats are very good examples. Pt, V_2O_5 and MnO_2 are used to catalyse the union of O_2 and SO_2 in the manufacture of H_2SO_4.

Large quantities of solid catalysts are used in petroleum industry for cracking to increase the yield of gasoline from petroleum.

2.10.1 Technical Importance of Industrial Catalysts

Some common catalysts, their importance and types of reactions are given below as:

Sl. No.	*Type of Reactions*	*Common Catalysts*	*Important Examples*
1.	Oxidation	(*i*) Spongy Pt, MnO_2, rare earth oxides	(*i*) Oxidation of NH_3 to HNO_3
		(*ii*) Pt, V_2O_5	(*ii*) Oxidation of SO_2 to SO_3
		(*iii*) Ag_2O	(*iii*) Preparation of ethylene oxide from ethylene
		(*iv*) Cu, Ag and their oxides	(*iv*) Oxidation of alcohols to aldehydes
2.	Hydrogenation	(*i*) Ni	(*i*) Hydrogenation of fats and oil
		(*ii*) Promoter Fe	(*ii*) Synthesis of ammonia
3.	Dehydrogen-ation	(*i*) Cr_2O_3, Al_2O_3	(*i*) Preparation of Styrene from ethyl benzene
4.	Hydration	(*i*) Hg salts	(*i*) Preparation of CH_3CHO from C_2H_2
5.	Dehydration	(*i*) H_3PO_4, H_2SO_4	(*i*) Preparation of amines from amides
		(*ii*) Al_2O_3, SiO_2	(*ii*) Preparation of olefines from alcohols
6.	Halogenation	(*i*) $FeCl_3$	(*i*) Preparation of Tetrachloroethane
7.	Polymerisation	(*i*) Oxides of transition metals and also Cu_2Cl_2, H_3PO_4, BF_3, $AlCl_3$ etc.	(*i*) Preparation of plastics from ethylene, propylene, phenols, aldehydes, venyl compounds etc.

Following are the desirable characteristics of an industrial catalysts:

(*i*) High surface area and porosity

(*ii*) Possibility of easy regeneration of reactivation

(*iii*) High thermal activity and mechanical strength

(*iv*) Long life

(*v*) Freedom from any constituent which may impair the catalyst activity.

2.10.2 Industrial Applications of Catalysts

Following are the main industrial applications of catalysts:

Sl. No.	*Process*	*Catalyst*	*Optimum Temperature*	*Reaction*	*Remarks*
1.	Haber's Process	Fe	450°C	$N_2 + 3H_2 \rightleftharpoons 2NH_3$	Used for the manufacture of ammonia. Mo & Al_2O_3 is used as promoters

2.	Contact process	V_2O_5 or Platinised asbestos	450°C	$2SO_2 + O_2 \rightleftharpoons 2SO_3$	Used for the manufacture of sulphuric acid
3.	Chamber process	NO_2	—	$2SO_2 + O_2 \rightleftharpoons 2SO_3$	Used for the manufacture of sulphuric acid
4.	Bosch's process	Fe_2O_3	400–450°C	$(CO + H_2) + H_2O \rightleftharpoons CO_2 + 2H_2$	Used for the manufacture of Hydrogen, Cr_2O_3 is used as promoter
5.	Vinegar process	Mycoderma-acetii	30–35°C	$2\ C_2H_5OH + O_2 \rightleftharpoons 2CH_3COOH + 2H_2O$	Used for the manufacture of acetic acid.
6.	Hardening of oils	Ni	150–300°C	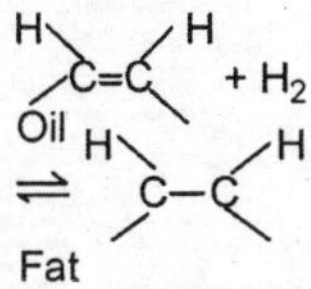	Used for the manufacture of vegetable ghee from vegetable oil

PROBLEMS

1. Write a brief note on the theories of catalysis.
2. Distinguish between:
   (*i*) Homogeneous and heterogeneous catalysis.
   (*ii*) Catalytic promoters and poisons.
   (*iii*) Positive and negative catalyst.
3. What do you mean by energy of activation? Explain with the help of energy reaction coordinate diagram. How does the nature of activated complex change in the presence of a catalyst.
4. What effect does a catalyst have on equilibrium system? Explain from the point of view of kinetics.
5. Explain the action of promoters and poisons with examples.
6. How does the nature of the activated complex change in the presence of a catalyst?
7. How does the concept of activation energy explain the role of catalyst in a chemical reaction?
8. Discuss the modern theory of heterogeneous catalysis.
9. How does a catalyst act in a reversible reaction in equilibrium? Give the reasons for your answer.
10. Discuss the mechanism of heterogeneous catalysis.
11. What are the characteristics of a catalytic reaction?
12. Discuss the intermediate compound formation theory of catalysis.
13. Define the term 'Catalysis', 'Inhibition' and 'Auto catalysis' with the help of examples.
14. Differentiate between homogeneous and heterogeneous catalysis with examples.

CHAPTER

3

AIR AND WATER POLLUTION

3.1 AIR POLLUTION

3.1.1 Introduction

Air pollution may be defined as any atmospheric condition in which certain substances are present in such concentrations that they can produce undesirable effects on man and his environment. These substances include gases such as sulphur oxides, nitrogen oxides, carbon monoxide, hydrocarbons etc., particulate matter such as smoke, dust, fumes and aerosols, radioactive materials and many others. Most of these substances are naturally present in the atmosphere in low (background) concentrations and are usually considered to be harmless.

A particular substance can be considered as an air pollutant only when its concentration is relatively high compared with the background value and causes adverse effects *e.g.*, SO_2, if present in the atmosphere in concentrations greater than the background value *i.e.*, 2×10^{-4} ppm and causes measurable effects on humans, animals, plants, or property, then only it is classified as an air pollutants.

Air constitutes about 80% of man's daily intake by weight. Human beings breathe nearly 22,000 times a day, inhaling about 16 kg of oxygen. Thus, pollution of air may have profound health effects and other consequences. The atmosphere is a mixture of about 5,500 million tones of gases, predominantly nitrogen (78.09%) and oxygen (20.94%). Other gases in the atmosphere together constitute less than 1%. These values essentially remain constant up to 50,000 altitude.

Other minor constituents in the atmosphere may include dusts, bacterial spores, smoke particles, H_2S and hydrocarbons depending on the climatic conditions, volcanic activity and distribution of animal, plant and human populations.

Pollution of the atmosphere indoors (inside enclosed buildings, residence, office etc.) is called indoor air pollution. Typical atmospheric pollutants encountered mainly indoors or outdoors, according to their sources are listed in Table 3.1. Human beings spend approximately

80% of their time indoors thus the degree of contamination of the indoor air is of considerable concern as well.

3.1.2 Types of Air Pollutants

The following are the major air pollutants:

(A) Gases

(i) Inorganics

Oxides of Carbon (CO, CO_2)

Oxides of Nitrogen (NO, NO_2)

Oxides of Sulphur (SO_2, SO_3)

Other inorganics (NH_4, HF, $H_2S_2O_8$)

(ii) Organics

Simple hydrocarbons

Aliphatics

Alkenes (ethence, propene)

Alkanes (ethane, methane)

Alkynes (acetylene)

Aromatics (benzene, toluene, xylene)

Oxygenated hydrocarbons

Aldehydes (formaldehyde, acetaldehyde)

Acids (formic acid, acetic acid)

Ketones (acetone)

Table 3.1: Typical sources of air pollutants grouped by origin

Pollutants	*Sources*
Group I. Sources predominantly outdoor:	
Lead	Automobiles, industrial emissions
Ozone	Photochemical reactions in the atmosphere
SO_2	Fuel combustion, smelters
Pollens	Trees and other plants
Organic Substances	Petrochemical solvents, natural sources, vaporization of unburnt fuels.
Group II. Sources both indoor and outdoor:	
CO_2	Metabolic activity, combustion
CO and nitrogen oxides	Fuel burning and tobacco smoke
Organic substances	Volatilization, combustion, metabolic action, pesticides
Polycyclic hydrocarbons	Fuel burning, tobacco smoke
Suspended particulate matter	Resuspension, condensation of vapours and combustion of products
Water vapour	Biological activity, combustion, evaporation

Group III. Sources predominantly indoor	
Aerosols	Consumer products
Allergens	House dusts
Ammonia	Metabolic activity, cleaning products
Asbestos and synthetic fibre	Fire retardant, acoustics, thermal or electrical insulation
Formaldehyde	Particle board, insulation, furnishings, tobacco smoke
Mercury	Spills in dental care facilities or laboratories, thermometer breakage
Organic substances	Adhesives, cooking, cosmetics, solvents
Radon	Building materials (Concrete, stone), water

(B) Particulates

(*i*) Solid particulates (dust, smoke)

(*ii*) Liquid particulates (fumes, mists)

The major factors contributing to air pollution in an industrial society are given in the Table 3.2.

Table 3.2: Air pollution in an industrial society

Emission Sources	SO_2	NO_2	*HC*	*CO*	*Particulates*
Motor vehicles	—	••	••	••	—
Industrial+	•	•	•	—	••
Power plants	••	•	—	—	•

\+ Pulp & paper, iron & steel, petroleum refineries, smelter, synthetic rubber and fertilizer, manufacturing units.

• This man-mad source is of considerable importance.

— This man-made source is of major importances.

Most, if not all, of the gaseous air pollutants after being emitted undergo various transformations in the atmosphere. The major process that coverts to the pollutants from their primary form at emission to other secondary forms are oxidation, dissociation and dissolution, as given in Table 3.3.

Table 3.3

Reactions	*Primary Pollutants*	*Secondary Pollutants*
Oxidation	Sulphur dioxide Hydrogen sulphide Nitric oxide Peroxyacyl nitrate (PAN) Carbon monoxide Certain organic radicals	Sulphur trioxide Sulphur dioxide Nitrogen dioxide Ozone Carbon dioxide Peroxy acylnitrate
Dissociation	Nitrogen dioxide	Nitric oxide + Oxygen

	Oxygen (molecular) Hydrocarbons & aldehydes	Oxygen (atomic) Organic radicals
Dissolution	Carbon dioxide Hydrogen fluoride Nitrogen dioxide Nitrous acid Sulphur dioxide Sulphur trioxide	Carbonic acid Hydrofluoric acid Nitrous/Nitric acid Nitric acid Sulphurous acid Sulphuric acid

The significance of the formation of secondary pollutants can be appreciated from the formation of ozone and peroxyacyl nitrates (PAN) from the aldehydes and hydrocarbon molecules.

3.1.3 Sources of Air Pollution

The sources of air pollutants are numerous. However, the main sources of air pollutants are as listed:

(*i*) Combustion,
(*ii*) Vaporization, and
(*iii*) Frictional force.

Combustion

Combustion is a chemical process involving the union of carbon and oxygen resulting in the production of heat and frequency of light energy as well. Such processes are represented by the simple equation:

$$C + O_2 \rightarrow CO_2 + \text{Energy}$$

A typical example could be burning of coal and coke (which are chiefly carbonaceous). Other processes involving a similar phenomenon are: domestic burning, agricultural refuse, firewood and dry dung burning. The Energy Survey Committee of India reported that about 12 million tonnes of wood, 50 million tonnes of dry dung and 30 million tonnes of vegetable wastes are burnt every year in India, largely in villages but also in urban areas.

It is the process of coal combustion that tends to be the main source of air pollution. The complete combustion of coal would result only in carbon dioxide (CO_2), an incomplete process, as it occurs in most cases, results in the production of CO and carbon particles alongwith CO_2. Also most of the fuels contain sulphur and nitrogen as minor contaminants, and this results in the release of oxides of sulphur and nitrogen during the combustion processes Table 3.4 provides an estimate of the pollutants emitted as a result of coal combustion processes.

Table 3.4: Gasesous pollutants emitted by coal combustion (per tonne of coal)

Pollutants	*Electricity Generating Plants*	*Industrial Plants*	*Domestic and Commercials Plants*
Sulphur dioxide (as SO_2)	19 S*	19 S*	19 S*
Nitrogen Oxide (as NO_2)	10	10	4
Carbon monoxide	0. 25	1. 5	25
Hydrocarbons (as CH_4)	0. 1	0. 5	5
Aldehydes (as HCHO)	0. 0025	0. 0025	0. 0025

Note: S* indicates that S in the coal should be multiplied by 19, *e.g*, if the S content is 2%, then the S emission would be 2 times 19, or 38 kg of SO_2 per tonne of coal.

Apart from these, coal may contain certain trace elements like Pb, Cd, Cr, Co, Cu, Mn, Ni and Zn, which provides, an additional source of heavy metal pollution. These trace elements present in coal get into volatile state in the gasifiers under reducing conditions.

Another potentially important fuel, used in the energy protection sector is petroleum (or oil). In recent years, petroleum derivatives have replaced coal as the principal source of energy in many places: the use of fuel oil for heating and electric power generation, gasoline for automobiles, kerosene in jet aeroplanes and diesel oil for automobiles and various appliances are typical examples.

Oils are the primary sources of highly dangerous hydrocarbons in the environment. Moreover, they have led to new type of air pollution problems in which photochemical reactions play an important role. Refining of oil, its combustion during use and losses during transportation are the major points, where the oil tends to pollute the environment.

Another major source of air-borne pollutants, through the combustion processes is the automobile. Pollution occurs each time a motorist presses his brake pedal. Thus vehicles on road serve as mobile sources of air pollution, contributing such pollutants as CO, NO_2, smoke, organic vapours and incompletely-burnt hydrocarbons as shown in Table 3.5.

Table 3.5: Average emission factors of pollution for automobile exhausts

Types of Emission	*kg/1000 litres of Gasoline used*
Carbon monoxide	300.00
Hydrocarbons	25.00
Oxides of Nitrogen	14.00
Particulates	1.50
Oxides of sulphur	1.00
Aldehydes	0.50
Organic acids (acetic)	0.50
Benzo(a) pyrene*	0.06

* g/1000 litre

Leaded gasoline is by far the most significant source of lead in the environment. Lead (as tetra alkyl lead) is added to gasoline to increase its octane-rating and facilitate better combustion by preventing pre-ignition losses.

Nickel may be emitted from the Ni-added to gasoline and Ni-containing parts of automobiles. Zinc and Cd come from the lubricating motor oils, tyres and galvanized parts of the vehicles.

Diesel vehicles are important sources of nitrogen oxides, hydrocarbons and particulates, as shown in the Table 3.6.

Table 3.6: Emission factors for diesel engines

Types of Emission	*Emission Levels (kg/1000 litres of diesel)*
Oxides of nitrogen	27
Hydrocarbons	17
Particulate matter	14
Carbon monoxide	07
Oxides of sulphur	05
Organic acids	04
Aldehydes	01
Benzo(a) pyrene*	0.1

* g/1000 litres.

Aircraft account only for about 1% of the total hydrocarbons, nitrogen oxides and carbon monoxide, and for an even smaller fraction of particulate matter and sulphur dioxides emitted.

The extent and importance of air pollution by combustion are obvious when one considers the amount of air consumed by these processes. Estimates show that 1 kg of gasoline requires 9 M^3 of air, 1 kg of oil requires 9.4 M^3, 1 kg of coal requires 10.5 M^3 and 1 kg of natural gas uses up 14 M^3 of air in the process of combustion. Such combustion processes consume and pollute about 65.5 KM^3 of air everyday in New Delhi, 81.5 KM^3 in Mumbai and 86 KM^3 in Kolkata.

Vaporization

Vaporization denotes a change in the physical state of a substance from liquid to gaseous form. A volatile liquid readily produces vapours when exposed to the atmosphere. This occurs when the solvents are being used in laboratories and industries. Industrial units associated with certain chemical products such as rubber, paints and plastics may, in general, produce vapourous by-products during the manufacturing processes. Many vapour pollutants also results from the spray painting of furnitures, automobiles and instruments. Moreover, agricultural activities such as pesticide spraying may also contribute in certain cases. Smoking of cigarettes releases polynuclear aromatic hydrocarbons (PAH), nitric and nitrous oxides (NO, N_2O) and carbon monoxide (CO). Vaporization may also be responsible for the entry of organic chemicals and hydrocarbons from the moving automobiles.

Frictional force

Air-borne particulate matter, mainly arises from the wear and tear of exposed surfaces, or frictional activities. Examples of this category would be, crushing, grinding, screening, sawing and drilling operations, mainly noticed in the vicinity of ore extraction, mining and construction industries.

3.1.4 Pollution Control Techniques

The most effective means of dealing with the problem of air pollution is to prevent the formation of the pollutants or minimize their emission at the source itself. In the case of industrial pollutants, this can often be achieved by investigating various approaches at an early stage of process design and development, and selecting those methods which do not contribute to air pollution or have the minimum air pollution potential. These are known as **source correction methods.** Application of these methods to existing plants is difficult, but still some of these correction methods could be applied without severely upsetting the economy of the operation. Control of the pollutants at the source can be accomplished in several ways through **raw material changes, modification or replacement of process equipment,** and by more effective operation of existing equipment.

When source correction methods cannot achieve the desired goal of air pollution control, use is made of effluent gas cleaning techniques. These involve many of the chemical engineering unit operations and at present form the main part of pollution control technology.

3.1.4.1 *Source Correction Methods*

(A) Raw material changes

If a particular raw material is responsible for causing air pollution, use of a pure grade of raw material is often beneficial and may reduce the formation of undesirable impurities and by-products or may even eliminate the troublesome effluent. A typical example of this approach is the use of low sulphur fuel in place of high-sulphur ones.

(B) Process change

Process changes involving new or modified techniques offer important ways of lowering atmospheric pollutant emissions. Radical changes in petroleum and chemical refining industries have resulted in minimizing the release of materials to the atmosphere. The volatile substances are recovered by condensation and the non-condensable gases are recycled for additional reactions. Hydrogen sulphide, which was once flared in refineries, is now recycled and used in Claus process to recover elemental sulphur.

Rotary kilns are a major source of dust generation in cement plants. Some degree of dust control may be achieved in the kiln operation by adjusting operating conditions.

Smelting and paper industries are notorious for emitting highly objectionable sulphurous materials. These emissions are being curtailed by major process changes such as hydro-metallurgical separations of ores and use of no sulphides in paper making.

In the steel industry, a radically different process has been proposed to lower sulphurous emission during combustion. In this process the sulphur bearing fuel, limestone and air are injected into a molten iron bath. The combustibles in the fuel are partially oxidized to carbon monoxide within the molten iron bath; the gaseous CO comes off at the top of the molten iron and is burnt efficiently in a conventional manner. The sulphur is retained in iron bath and forms a slag with the limestone which removed.

(C) Equipment modification or replacement

Air pollutant emission can be minimised by suitable modifications or replacement of process equipment *e.g.*, the unburnt CO and hydrocarbons in the cylinders of an automobile engine, which are otherwise emitted into the atmosphere through the tail pipe, can be burnt by injecting air into the hot exhaust manifold of the engine. Similar results can be obtained by suitable modifications in the carburator and ignition systems.

In petroleum refineries, hydrocarbon vapours are released into the atmosphere from storage tanks due to temperature changes, direct evaporation; and displacement during filling. These losses can be minimized by designing the tanks with floating roof covers or by pressuring the tanks.

In addition to the above mentioned source correction methods, air pollutant emissions from industrial operations can be reduced by proper equipment maintenance, housekeeping and cleanliness in the facilities and premises. Often changes in the design of local exhausthood and proper installation can minimize the emission of pollutants to the atmosphere.

Chemical process plants often have extensive leakage around ducts, piping, valves and pumps. Many such leaks can be prevented by checking the seals and gaskets routinely.

3.1.4.2 *Cleaning of Gaseous Pollutants*

The technology for the removal of gaseous pollutant emissions after their formation has probably received the maximum attention. The cleaning techniques are applied to those cases where emission of pollutants cannot be prevented and pollution control equiplents is necessary to remove them from the main gas stream. Normally, it is more economical to install the control equipment at the source where the pollutants are present in the smallest possible volume in relatively high concentrations rather than at some point away from the source where the pollutants are diluted by other process gases or air.

Among the gaseous air pollutants, the carbon dioxide (CO_2), carbon monoxide (CO), sulphur dioxide (SO_2), oxides of nitrogen (NO, NO_2) and hydrocarbons have attracted major attention because of their common occurrence in the atmosphere, and their known harmful effects.

(A) Control of carbon-dioxide

The emission of CO_2 needs to the controlled in view of the possible onset of a disastrous global warming process. There are three possible solutions to the problems created by the CO_2 emissions. They are as follows:

(*i*) Reduction in the use of the fossil fuels.

(*ii*) Disposal of the CO_2 liberated into the atmosphere.

(*iii*) Controlling the entry of CO_2 into the atmosphere, by collecting it at the point of entry and disposing it off safely elsewhere.

The only way of disposal that is now and always working is the removal of CO_2 by forests and other green vegetations. Since the plants take up CO_2 from their synthetic processes, they seem to be the excellent solution to this problem. However, with the extent of deforestation that has been done, the problem is enlarging in magnitude day by day. Scientists suggest that an area of the size of France should be planted afresh every year to compensate for the current rate at which CO_2 is being produced.

(B) Control of ozone

Action plans for the control of ozone-depletion are underway in several countries. The United States, *e.g.,* has banned the use of CFCs (chlorofluorocarbons, that destroy the stratospheric ozone) in 1978. Also several of the EEC (European Economic Community) nations have banned CFC production altogether. A protocol for the protection of the ozone layer was adopted in Vienna in March 1985 by 21st century and the EEC. The convention pledged to protect human health and the environment from the effects of ozone depletion.

(C) Control of sulphur-dioxide

The greatest anthropogenic sources of SO_2 emissions are the fossil fuel burning power plant. There are three basic procedures for controlling sulphur dioxide emissions from stationary combustion sources:

(*i*) the extraction of sulphur from fuels,

(*ii*) sulphur reduction within the combustion chamber, and

(*iii*) the treatment of fuel gases.

Sulphur dioxide emissions can be reduced by switching to low sulphur fuels, doing fuel de-sulphurication, adopting alternative energy sources (*e.g.*, nuclear energy) or by applying control either before, after or during the combustion process.

Sulphur in coal is present in both inorganic and organic forms. Practically all inorganic sulphur exists in the coal as iron disulphide (FeS_2), *i.e.*, in the form of pyrites and marcasites. Washing can reduce the pyritic sulphur content by about 30%, but such methods result in the loss of combustible material and may add 20% or more to the cost of coal. Organic sulphur, present in the form of cystin, thiols, sulphides and some cyclic compounds, is intrinsically and almost uniformly bound with most of the carbonaceous material and can be removed only by chemical processing. The desulphurication is the process of physical removal of sulphur from the fuels. The success of this process depends on the amount of pyritic sulphur present in the coal.

The option that is attracting much attention is the removal of SO_2 during the process of combustion. An example of this approach is the addition of dolomite or limestone ($CaCO_3$) in the combustion chamber, readily converts the SO_2 liberated into solid $CaSO_4$, which can readily be removed alongwith the ash.

$$CaCO_3 \rightarrow CaO + CO_2$$

$$CaO + SO_2 \rightarrow CaSO_3$$

$$CaO + SO_2 + 1/2O_2 \rightarrow CaSO_4$$

Substitution of energy sources, a potential alternative, has its limits. Hydroelectric plants require no fuel and certainly are free of SO_x pollution. The use of nuclear power to generate electricity could virtually eliminate SO_x pollution from the power plants.

(D) Control of carbon monoxide

About two-thirds of the CO emitted are from the gasoline-fueled internal combustion engines. Since changing the options to reduce CO emissions in the internal combustion engines affects the emission levels of other contaminants, (hydrocarbons, nitrogen oxides and particulates) as well, it is a difficult one. When the air-fuel ratio is lowered, the NO_x emission is minimized but the CO and HC emission increases. Though the apparent solutions seem to be adopting a high air-fuel ratio, therefore, the engine design becomes more difficult. However, the basic principle in the control of CO emission is to convert CO to CO_2. The enable such conversions, catalytic reactors are being equipped outside the engines. The problems of increasing CO_2 has recently been recognized and that warrants pollution control measures as well.

(E) Control of nitrogen oxide

The control of nitrogen oxide emissions has mainly been focussed on two things:

(*i*) Modification of the combustion conditions as a means of decreasing the amount of NO_x liberated, and

(*ii*) Utilization of various devices to remove, NO_x from the exhaust gases.

Catalytic reduction of NO_x to N_2 has been attempted particularly as a solution to the problem of NO_x emission from the automobiles.

NO_x is removed by passing the gas through H_2SO_4 solution or alkaline scrubbing solutions containing $Ca(OH)_2$ and $Mg(OH)_2$. Hence, in this process NO is converted into N_2O_3 which is easily absorbed.

$$NO_2 + NO \rightarrow N_2O_3$$

Normally, four steps are involved in the above scrubbing process:

Step I: NO_2 and fuel gas are introduced into an oxidizer

$$NO_2 + SO_2 + H_2O \rightarrow H_2SO_4 + NO$$

Step II: NO_2 and NO react to form N_2O_3, which is scrubbed by H_2SO_4.

$$NO_2 + NO \rightarrow N_2O_3$$

$$N_2O_3 + 2H_2SO_4 \rightarrow 2NOHSO_4 + H_2O$$

The cleaned fuel gas is released into the atmosphere.

Step III: The product of IInd steps *i.e.*, $NOHSO_4$ is decomposed alongwith the evolution of H_2SO_4

$$NOHSO_4 + \frac{1}{2}O_2 + H_2O \rightarrow H_2SO_4 + 2NO_2$$

Step IV: NO_2 reacts with H_2O in the HNO_3 reactor resulting in the production of HNO_3

$$3NO_2 + H_2O \rightarrow 2HNO_3 + NO$$

NO produced in the IVth step and excess NO_2 are recirculated through the oxidizer in Step I.

(F) Control of hydrocarbons

Four techniques are used to control the hydrocarbon (HC) emissions from stationary sources. These techniques are:

(*i*) incineration,

(*ii*) adsorption,

(*iii*) absorption, and

(*iv*) condensation.

The incineration aims at the conversion of the hydrocarbons to CO_2 and water. In adsorption, the hydrocarbon vapours are adsorbed onto the surface of carbon and remain there until they are periodically removed by the passage of stream through the system. The absorption method is similar to the adsorption except that the exhaust is brought into intimate contact with a liquid in which the hydrocarbons will dissolve or become suspended, which can latter be removed. The hydrocarbons can as well be condensed and removed as such. But in the mobile sources the general aim has been to reduce the HC emissions by completing the process of combustion and enabling their complete conversion into carbon dioxide (CO_2), for example.

(G) Control of particulate matter

Techniques for controlling particulate emissions are aimed at preventing their entry into the atmosphere. The logical basis of controls is to capture the particulates as they are emitted. The following are the different types of equipment in use:

(*i*) Cyclone collectors

(*ii*) Wet scrubbers

(*iii*) Electrostatic precipitators

(*iv*) Gravity setting chamber

(*i*) Cyclone collectors: These equipment consist of a tight circular spiral that produces a centrifugal force, as a result of which the suspended particles, in the following gas are forced to move outward to a wall where they are collected. This process removes all small particulates (diameter 5-20 microns), efficiently.

(*ii*) Wet scrubbers: In the scrubbers of this type a liquid is utilized to assist in the removal of solid, liquid or gaseous contaminants. The effectiveness of these devices depends primarily upon the frequency of contact and the degree of interaction between the liquid phase and the contaminant to be removed.

(*iii*) Electrostatic precipitators: The electrostatic precipitators (tube or plate type) are used to remove the particulate matter during their passage through a region of high electrical charge. It is during this process that the moving particulates become charged and are then attracted to an oppositely charged area, where they are removed effectively.

(*iv*) Gravity setting chambers: These are large chambers which allow the velocities of the gaseous matter to decrease and thereby let the dusts or droplets to settle out. Large particulates (diameter > 50 microns) in the emission atmosphere are effectively removed in this way. The settling time required for smaller particles is so long that the use of this type of an equipment is impractical in such cases.

3.1.4.3 *Suggestions for Control of Air Pollution*

(*i*) Emissions from automobiles and vehicles may be minimized by cleaning the exhaust gases after the combustion through the use of catalyst in catalytic converters.

(*ii*) Smoke can be reduced during combustion of fuel by correct method of firing admitting correct quantity of air, maintaining high temperature and continuous feeding of fuel in place of intermittent feeding.

(*iii*) Industrial smoke can be reduced by installing electrostatic precipitator.

(*iv*) SO_2 can be controlled either by removing it after combustion or by pre-combustion or by de-sulfurication of fuel.

(*v*) Residential colonies should be built far away from the industries.

(*vi*) Vehicles with old technology should be banned from operating on roads and taxed heavily.

(*vii*) For the elimination of dust, extraction, ventilation can be applied. Other methods for removing dust are bag filters, cyclone dust separators, centrifugal separators, washers, etc.

(*viii*) Strict legislations for the use of tall stacks or chimneys as these can bring substantial reduction in the concentration of air pollutants at the ground level and provisions for industrial filters for the exhaust that is produced in the thermal power stations and other industries using fuel as energy sources.

(*ix*) Heavy duty vehicles should not be allowed to ply on the inner city roads at peak hours. This would improve the ppm figure of the suspended particles inside the city.

(*x*) Introduction of cleaner emission technology in vehicles should be encouraged and the price difference between diesel and petrol should be minimized as diesel is a greater source of pollutant release.

(*xi*) Existing knowledge of air pollution/health damage relationship may be improved, extended and evaluated to full range to resolve the problems of air pollution.

3.1.5 Effect of Air Pollutants on Human Beings

Air pollution, as it is related to health, is basically an ecological problem. It is a problem faced not only by human beings but also by other forms of animal as well as plant life. Most air pollutants, like other forms of matter, in time, appear in water, or soil, and eventually in human food. The effects of air pollutants on human beings is summarized in Table 3.7 and 3.8.

Table 3.7: Effects of gaseous pollutants

Pollutants	*Effects*
Sulphur dioxide (SO_2)	Breathing rate increases while breathing itself becomes less deep and a general feeling of air starvation is experienced. Some even show signs of *Bronchial spasms* and asthmatics are badly affected. Any SO_2 which oxidizes to SO_3 may dissolve in body fluids to form H_2SO_4, a very strong and corrosive acid.
Carbon monoxide (CO)	CO combines with the haemoglobin in the blood to form carboxy haemoglobin and when about half of the haemoglobin of the blood is used up, death ensures as the carboxy haemoglobin is useless for respiratory purposes. Persons poisoned by carbon monoxide exhibit a characteristic bright pink colour of the flesh due to the presence of carboxy haemoglobin in the blood.
Oxides of nitrogen (NO_x)	Prolonged exposure of human beings to even low concentrations of oxides of nitrogen, can cause a chronic irritation of the respiratory tract, headache, loss of appetite and corrosion of teeth. Nitric oxides reacts with air to form brown nitrogen dioxide which is very poisonous.
Ozone (O_3)	In high concentrations, ozone exhibits a noticeably blue colour and an odour not unlike chlorine. It is poisonous in comparatively low concentrations and can cause irritation and a total ozone concentration of 26 g/m^3 may lead to asthma attacks.

Table 3.8: Effects of non-biological air pollutants on human beings

Pollutants	*Sources*	*Effects*
Aldehyde (formaldehyde and acrolein)	Autoexhaust, waste incineration, fuel combustion, photo-chemical reactions.	Odours irritant to eye, skin and respiratory system.
Ammonia	Chemical industries, coke ovens, refineries, stocks yards and fuel incineration.	Corrosive to mucous membranes, damaging to eyes and respiratory tract.
Arsenic	Metal smelters, arsenical pesticides and herbicides.	If inhaled, ingested to absorbed, through skin. It causes dermatitis, mild bronchitis and nasal irritation, suspected carcinogen.
Asbestos	Asbestos factory or mine construction sites.	Pulmonary fibrosis, pleural calcification, lung cancer.
Barium	Industries, mining, refining, barium based chemicals, smoke suppressant, additives in diesel fuels.	Affects heart muscles, gas trointestinal tract and central nervous system and respiratory tract.
Beryllium	Industrial usage, production of fluorescent lamps, motor fuels.	Pulmonary damage and damage to skin membranes from handling and mucous soluble salts of beryllium.
Boron	Industry producing boron, petroleum fuel, or additives, present in coal.	Toxic through ingestion of inhalation, as dust causes irritation and inflammation; boron hydrides can cause damage to central nervous system which may result in death.
Cadmium	Metal industries engaged in extraction, refining, machining, electroplating and welding materials. By-product of refining lead, zinc and copper. Pesticides, fertilizer, cadmium, nickel batteries, reactor poison in nuclear fission plants, used in production of tetraethyl lead gasoline.	Chronic and acute poisoning. Inhalation of fumes and vapours, causes damage to kidney, emphysema, bronchitis, cancer, gastric and intestinal disorders, diseases of heat, liver and brain.
Chlorine	Process industries using chlorine, accidental breakage during storage and transportation.	Irritates eye, nose and throat. Large doses damage lungs, producing edema, pneumonitis, emphysema and bronchitis.

Chromium	Metallurgical and chemical industries, products employing chromate compounds, cement, asbestos.	Irritative, corrosive and toxic to body tissues. Believed to cause carcinogenic action. Develops dermatitis and ulcers of the skin and perforation of nasal septum.
Ethylene	Motor vehicle emissions, chemical industries, incineration of agricultural wastes, emitted by growing plants.	Eye irritation as a result of photo-chemical reaction with nitrogen oxide and ozone.
Hydrochloric acid	Byproduct from chlorination of organic compounds, burning of chlorinated plastics and paper, combustion of gasoline containing ethylene chloride.	Coughing and choking from inhalation, inflammation and ulceration of upper respiratory tract, clouding of the cornea.
Hydrogen sulphide	Biological decay of protein material in stagnant water, craft paper mills, industrial waste disposal ponds, sewage treatment plant, refineries, coke ovens.	Headaches, connectivities, sleep-lessness, and pain in eyes. In high concentrations, can lead to blockage of O_2 transfer, it acts as a cell poison and damages nerve tissues.
Iron	Iron and steel plants, fly ash from combustion of coal and fuel oil, municipal waste incineration, use of welding rods.	Benign pneumoconiosis and siderosis (iron pigmentation of the lungs). Iron oxides act as a vehicle for transporting carcinogens and sulphur dioxide deep into the lungs.
Lead	Automobile emissions, lead smelters, combustion of coal and fuel oil, lead arsenate pesticides.	Absorbed through gastro-intestinal and respiratory tract and deposited in mucous membrane of nose, throat and in the lungs.
Manganese	Blast furnaces producing ferromanganese compounds, organic manganese fuel additives, use of welding rods, incineration of manganese containing products.	Manganese poisoning of central nervous system, manganese pneumonia caused by absorption, inhalation, ingestion or through skin.
Mercury	Mining and refining of mercury, use of mercury in laboratories. Pesticides containing mercury.	Vapours inhaled may cause intoxication or protoplasmic poisoning.
Nickel	Metallurgical plants using nickel, engines burning fuels containing nickel additives, burning coal and oil, nickel plating facilities, incineration of nickel products.	May cause cancer of lungs and sinus, other respiratory disorders, dermatitis.

Phosphorus	Plants producing phosphate fertilizers, phosphoric acid, phosphorous peroxide. Emission from vehicles and aeroplanes using phosphorus as corrosion inhibitors in fuel.	Skin irritation to systematic poisoning. High concentrations affect nervous system.
Radioactive substances	Direct contamination by radioactive gases or suspended dust from natural or artificial sources. Indirect contamination when radionuclides are ingested from food chain as a result of contact with contaminated ground, water, plants or animals.	Somatic effects such as leukemia and other types of cancer, cataracts and reduction in life expectancy. Genetic effects include mutation in human gametes that show up only in future generations.
Selenium	Combustion of industrial and residual fuels, refinery waste gases and fumes, incineration of waste including paper products.	Irritation to eyes, nose, throat, respiratory tract, and gastro-intestinal tract, chronic intoxication from industrial toxicants. Effects on kidney, liver and lungs.
Vanadium	Vanadium refining industries, alloy industries, power plants and utilities using vanadium-rich residual oil.	Physiological effect of varying severity on the gastrointestinal and respiratory tracts. Inhibition of cholesterol synthesis. Chronic exposure related to heart disease and cancer.
Zinc	Zinc refineries, manufacture of brass, zinc-galvanizing process.	Fumes have a corrosive effect on the skin which irritate and damage the mucous membrane.

3.1.6 Chemical and Photochemical Reactions in the Atmosphere

Many of the gaseous pollutants undergo chemical reactions within the atmosphere and form either new compounds or aerosols. This mode or removal is of great importance for sulphur dioxide. A large part of SO_2 in the atmosphere is oxidized to SO_3, which quickly combines with moisture to form H_2SO_4 mist. The overall reaction is as follows:

$$2SO_2 + 2H_2O + O_2 \rightarrow 2H_2SO_4$$

The above reaction has been catalysed by metal salts such as iron and manganese, commonly formed in the flyash. These particles serve as nucleation sites for droplet formation and the sulphuric acid droplet may in turn react with metal salts (NaCl), metal oxides (MgO, Fe_2O_3, ZnO and Mn_2O_3) or ammonia to produce sulphates:

$$2NaCl + H_2SO_4 \rightarrow Na_2SO_4 + 2HCl$$

$$MgO + H_2SO_4 \rightarrow MgSO_4 + H_2O$$

$$2NH_3 + H_2SO_4 \rightarrow (NH_4)_2\,SO_4$$

With these reactions, the sulphuric acid droplet is neutralized and the solubility of SO_2 in the droplets is further increased thereby enhancing the oxidation process. Both the acid droplets and the sulphate particles are rapidly removed from the atmosphere by wet precipitation.

Similarly, nitrogen oxide is removed from the atmosphere through the conversion of NO_2 into HNO_3. The overall reaction is as follows:

$$2NO_2 + H_2O \rightleftharpoons HNO_3 + HNO_2$$

$$3HNO_2 \rightarrow HNO_3 + 2NO + H_2O$$

or

$$3NO_2 + H_2O \rightarrow 2HNO_3 + NO$$

Atmospheric chemical reactions usually require more activation energy than can be supplied by molecular collisions due to ambient air temperature. Because of this, the conversion of SO_2 to SO_3 and NO_2 to HNO_3 is very slow. Thus, the necessary activation energy for most atmospheric reactions is supplied by solar radiations, therefore, called photochemical reaction.

A photochemical reaction begins with the act of absorption of radiation. According to the Einstein law of photochemical equivalence, this is a quantum process involving one photon per absorbing molecules. The number of absorbing molecules is, therefore, equal to the number of photon absorbed. The immediate result of the absorption of a photon is an excited state of the absorbing molecule which has excess absorbed energy. Though, most electronic spectra lie in the visible and ultraviolet region, the ultraviolet region is of photochemical importance.

The electronically excited molecules produced by absorption may undergo any of the following changes:

(*i*) They may dissociate.
(*ii*) They may react with other molecules on collision.
(*iii*) They may internally rearrange or polymerize.
(*iv*) They may lose their excitation energy by fluorescence or deactivation and thereby return to their original state.

Any of these possibilities except the last one serves as an initial chemical step or primary process in a photochemical reaction. The three steps involved in an overall photochemical reaction are:

(*a*) Absorption of radiation,
(*b*) Primary reactions, and
(*c*) Secondary reactions.

Absorption of Radiation

The sunlight in the wavelength range below 2900 Å is removed effectively by the stratosphere ozone layer before reaching the lower atmosphere. It is in the ultraviolet region with a wavelength of 3000 to 5000 Å, that the radiation is of an use in photochemical reactions.

The ultraviolet light available in the troposphere contains sufficient energy to break carbon-carbon and nitrogen-oxygen bonds by primary photochemical reactions.

Light absorption of pollutant molecules may lead to the formation of an excited state in which the excited molecules interact. Alternatively, excited molecules may transfer their energy to other molecules or atoms or may lose their energy by emission of light as fluorescence.

The absorbers of solar radiation include the following components: NO_2, SO_2, aldehydes, nitrogen & HNO_3, H_2O_2, ketones, organic oxides, many other organic compounds and aerosols such as metal oxides.

Primary Reactions

The most important primary photochemical reaction is the dissociation of NO_2 into NO and atomic oxygen (O):

$$\underset{\text{Nitrogen dioxide}}{NO_2} + h\nu \rightarrow \underset{\text{Nitric oxide}}{NO} + O$$

Nitric oxide in the atmosphere is oxidized relatively slowly by oxygen or very rapidly by ozone to form NO_2. Consequently, the above reaction forms a reacting chain which may be broken only by the complete conversion of NO_2 to HNO_3 by hydration and catalytic oxidation in the presence of fog droplets and aerosols.

The hypothetical sequence of reactions may be as illustrated:

$$2NO + O_2 \rightarrow 2NO_2$$
$$NO + O_3 \rightarrow NO_2 + O_2$$
$$2NO_2 + O_3 \rightarrow N_2O_5 + O_2$$
$$4NO_2 + 2H_2O + O_2 \rightarrow 4HNO_3$$

Nitric acid and nitrous acid may absorb light and be dissociated photochemically to NO_2 and NO as follows:

$$HNO_3 + h\nu \rightarrow NO_2 + HO^\bullet$$
$$HNO_2 + h\nu \rightarrow NO + HO^\bullet$$
$$HNO_2 + h\nu \rightarrow NO_2 + H^\bullet$$

Similarly, sulphur dioxide in the presence of oxygen is slowly photolyzed when exposed to solar radiation, yielding sulphur trioxide (SO_3), and in the presence of water, to H_2SO_4.

$$SO_2 + h\nu \rightarrow SO_2^\bullet \text{ (excited)}$$
$$SO_2^\bullet + O_2 \rightarrow SO_4^\bullet$$
$$SO_4^\bullet + O_2 \rightarrow SO_3 + O_3$$
$$SO_3 + H_2O \rightarrow H_2SO_4$$

The rate of photo-oxidation of SO_2 and formation of H_2SO_4 aerosol greatly enhanced by the presence of olefin and the oxides of nitrogen.

Secondary Reactions

Secondary reactions are thermal and are independent of the means by which the reactants concerned are produced. The following example illustrates the formation of free radicals by the reaction of O_3 with olefins or reaction of free radicals with oxygen, ozone, NO and other organic substances.

$$RCH = CHR + O_2 \rightarrow 2RCHO + O_2$$
$$R^\bullet + O_2 \rightarrow ROO^\bullet$$
$$ROO^\bullet + NO^\bullet \rightarrow RCOONO$$

The alkyl and aryl radicals unite with oxygen to form peroxyl, a radical, which in turn reacts with oxygen to give O_2 and with NO_3 giving peroxyacyl nitrate (PAN), formaldehyde (an irritant) and polymeric compounds (which reduce visibility). The formation of PAN takes place as follows:

$$R\overset{\overset{O}{\|}}{C}OO^{\bullet} + NO_2 \rightarrow \underset{\text{(PAN)}}{R\overset{\overset{O}{\|}}{C}OONO_2}$$

PAN belongs to a family of compounds which includes peroxyacetyl nitrate, peroxypropionyl nitrate, etc.

3.1.7 Smog Formation

Photochemical smog was first observed in Los Angeles, USA in 1940 and since then the phenomenon has been detected in most metropolitan cities of the world. The conditions for the formation of photochemical smog are air stagnation, abundant sunlight and high concentrations of hydrocarbon and nitrogen oxides in the atmosphere.

Smog arises from photochemical reaction in the lower atmosphere by the interaction of hydrocarbons and nitrogen oxide released by exhausts of automobiles and some stationary sources. This interaction results in a series of complex reactions producing secondary pollutants such as ozone, aldehydes ketones and peroxyacyl nitrates.

These reactions are extremely complex in nature. The oxidant formed in largest concentration is ozone which is a serious pollutant in lower atmosphere. Higher up in the atmosphere there are sufficient short wave length UV radiations (1700 Å to 2400 Å) available to convert oxygen to ozone. However, these wavelengths do not reach Earth's surface. The excessive build up of ozone in lower atmosphere has been through to be due to the presence of nitrogen oxide which absorbs UV radiations of a little higher wavelengths (3000 Å to 4000 Å) and produces an atom of oxygen. This atom reacts with molecular oxygen to produce ozone. In unpolluted atmosphere ozone thus produced reacts with nitric oxide, also a product of above reaction and results in the formation of nitrogen dioxide and molecular oxygen again as shown in Figure. 3.1.

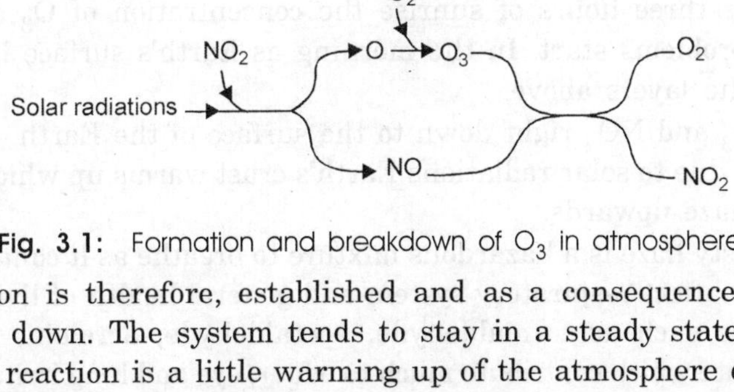

Fig. 3.1: Formation and breakdown of O_3 in atmosphere

A cyclic reaction is therefore, established and as a consequence most of the ozone produced is broken down. The system tends to stay in a steady state condition. The net result of this cyclic reaction is a little warming up of the atmosphere due to absorption of energy of solar radiations. The presence of hydrocarbons, especially the unsaturated ones (olefins), however, upset the natural balance. The oxygen atom produced by NO_2 reacts with hydrocarbons to covert them to their highly reactive oxidized state which starts reacting with nitric oxide to form NO_2 again, as shown in Figure. 3.2.

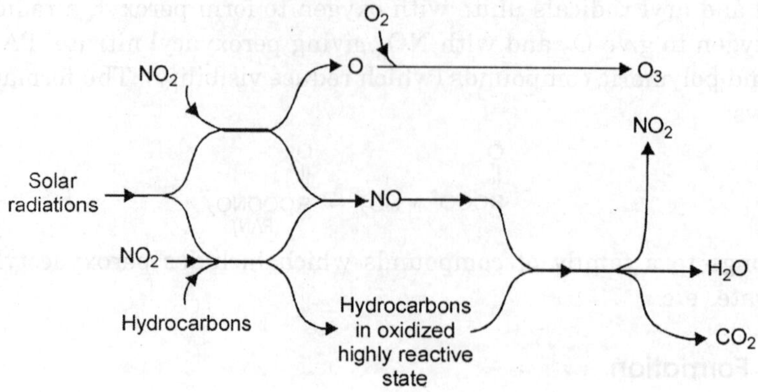

Fig. 3.2: Accumulation of O_3 and NO_2 in the atmosphere

Due to the greater affinity, nitric oxide present in the atmosphere reacts preferentially with reactive hydrocarbons and ozone is left out. As a consequence a gradual build up of the concentrations of NO_2 and O_3 occurs while that of NO declines. The brownish colour of the smog is due to the colour of NO_2 which is orangish brown.

The entire chain of these chemical reactions is initiated by absorption of solar radiations. Early in the morning the concentration of oxidants, NO_2 and O_3 are almost negligible while that of NO and hydrocarbons (HC) are a little higher. With the rise of life activity and morning traffic the concentrations of NO and HC begins to rise. As the Sun comes up NO_2 absorbs solar radiation splits up to yield oxygen atom which produce O_3 on reacting with O_2 and also cause the oxidation of hydrocarbon.

$$NO_2 + h\nu \rightarrow NO + O$$
$$O + O_2 \rightarrow O_3$$
$$HC + O \rightarrow RCO^{\bullet} \text{ (acyl radical)}$$

The concentration of O_3 begins to rise more rapidly as the Sun comes in while hydrocarbons are converted to their reactive state which triggers the reaction in which NO is rapidly changed to NO_2.

$$O_3 + NO \rightarrow NO_2 + O_2$$

Within two or three hours of sunrise the concentration of O_3 and NO_2 reaches the threshold where problems start. In the morning as Earth's surface is cooler, the air near it is cooler than the layers above.

This brings O_3 and NO_2 right down to the surface of the Earth. As the day advances, by 1.00 to 2.00 PM due to solar radiations Earth's crust warms up which gradually displaces the brown misty haze upwards.

The brown misty haze is a hazardous mixture to breathe as it contains O_2 and NO_2 both of which adversely affect respiratory system and general health of living organisms. It also contains aldehydes such as formaldehyde, acetaldehyde, acrolein, peroxyacetyl nitrate, peroxybenzoyl nitrate which contribute much to odour, eye and lung irritation and lacrimations caused by the photochemical smog. In plants prolonged exposures damage chloroplasts, disturb electron transport system and adversely affect a number of enzymes which play a vital role in photosynthesis. The formation of photochemical smog is a dynamic process which nature is illustrated in Figure. 3.3.

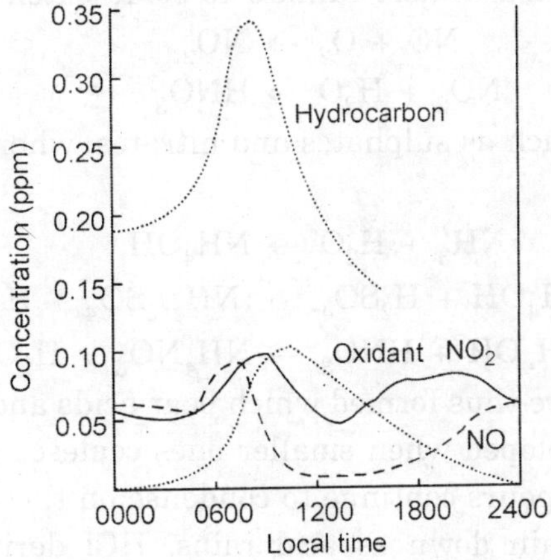

Fig. 3.3: Dynamic behaviour of photochemical smog

Los Angeles smog

Studies made in the Los Angeles area of the USA reveal an interesting sequence of concentration levels of various atmospheric pollutants during photo oxidation reactions that accompany photochemical smog episodes. It is initiated by the photo-dissociation of NO_2 and subsequent oxidation reactions involving unsaturated hydrocarbons, other organic compounds and free radicals to form O_3 and organic peroxides. The phenomenon occurs during warm, sunny days with gentle winds and low level inversion. Visibility is reduced markedly. Eye irritation, plant damage, a characteristic odour and accelerated cracking of rubber goods are obvious results.

An important feature of photochemical smog is the production of aerosols that markedly reduce visibility.

3.1.8 Acid Rain

When the pH of rain water or snow is less than 5.7, it is called acid rain. The phenomenon of acid rains is a consequence of accumulation of huge amounts of oxides of sulphur, nitrogen and fine particulate materials or aerosols. Water vapours condense on aerosol surface and form a fine film providing suitable loci as well as catalysts for the oxides of sulphur and nitrogen to dissolve in water and react to form to corresponding acids, i.e., H_2SO_3, H_2SO_4 and HNO_3.

Air-borne sulphur oxides (SO_2, SO_3) produced naturally or from human activities (such as the combustion of fossil fuels, etc.) can be absorbed by rainwater and snow, forming H_2SO_3 and H_2SO_4 respectively:

$$SO_2 + H_2O \rightarrow H_2SO_3$$
$$SO_3 + H_2O \rightarrow H_2SO_4$$

These acids (sulphurous and sulphuric) tend to lower the pH of the precipitation. The acid concentration of such rain water may range upto 60-70 μg/L. The acidity of the rainfall is largely a reflection of the amount of sulphurous materials (SO_2, SO_3 and SO_4^{--}) leached from the atmosphere.

Nitrates are also implicated in acid rainfall to some extent.

$$NO + O_2 \rightarrow NO_2$$
$$NO_2 + H_2O \rightarrow HNO_3$$

These acids form salts, such as sulphates and nitrates, when they come in contact with aerosols of basic nature.

$$NH_3 + H_2O \rightarrow NH_4OH$$
$$NH_4OH + H_2SO_4 \rightarrow (NH_4)_2SO_4 + H_2O$$
$$NH_4OH + HNO_3 \rightarrow NH_4NO_3 + H_2O$$

The secondary aerosols are thus formed which bear acids and salts of these acids. Larg and heavier aerosols are developed when smaller ones coalesce together and these tend to drift downwards. As water vapours continue to condense on these aerosols large droplets of water are produced which rain down as acid rains. HCl derived from the oxidation of organochlorine compounds and polyvinyl chloride (PVC) in plastics may also be present in these droplets.

Rain also traps much of wind blown particles, dusts and other gases constituents which may also be acidic in nature as it falls down.

Currently about 60–70% of acidic rain fall is caused by sulphur compounds and about 30–40% by nitrates. The oxidation processes (*i.e.*, nitrogen, sulphuric oxides to nitric, sulphuric acids) occur only at higher relative humidities (> 70%).

Consequences of acid rain

The damages caused by acid precipitation are of a very diverse nature and can be summed up as follows:

(*i*) *Damage of animals:* Most aquatic animals cannot survive a drop in pH, especially when it is lower than 4. Some species of fish, such as salmon die even when the pH drops to 5.5. Generally, at pH < 6, several susceptible species of algae and zooplankton are eliminated. A reduction in the zooplankton and bottom fauna will ultimately affect the food availability for the fish population.

(*ii*) *Damage of plants:* Corrosive nature of the acidic water itself is dangerous to plants. It decolourizes the leaf pigments, thus rendering them chlorophyl-less, result in decrease in agricultural productivity. Alphalpha, barley, cotton, lettuce and spinach are the non-woody plants most susceptible to acid-rain damage.

(*iii*) *Chemical imbalance:* The normal growth and development of plant and animal organisms are favoured in balanced ecosystems, where an equilibrium exists between available and non-available forms of an element. Acidic rainfall may upset such a chemical balance. Since the acid rain water flushing through the atmosphere may dissolve some metallic air pollutants, it can enhance their toxic action more directly, immediately and drastically, thus posing a serious threat to plant or animal life. The integrity of nutrient forms in various environmental compartments is altered.

(*iv*) *Material damage:* Sulphuric acid is highly reactive and corrosive. Metallic surfaces exposed to acid rain are susceptible to corrosion damage. Thus the metals, with which H_2SO_4 reacts, form their respective sulphates, *e.g.*, blue $FeSO_4$ results from iron, green

$CuSO_4$ from copper and white $Al_2(SO_4)_3$ from aluminium. On contact with acid rain water, textile fabrics, paper and leather products may lose their materials strength, get discoloured or simply disintegrate depending on the degree of acidity.

Building materials (limestone, marble, dolomite, mortor and slate) are especially weakened on reaction with acidic waters. Water, whose pH is acidic, can react with these water-insoluble carbonates to yield water-soluble sulphates which can easily be removed from the building structures:

$$CaCO_3 + H_2SO_4 \rightarrow CaSO_4 + H_2O + CO_2$$

$$CaMg(CO_3)_2 + 2H_2SO_4 \rightarrow CaSO_4 + MgSO_4 + 2H_2O + 2CO_2$$

This results in the pitting of the external surfaces and weakening of the building structure.

Many of our monuments which are built up of steel, stone of cement, concrete are threatened by the corrosive action of acid rains like the **Taj Mahal at Agra.**

3.1.9 Green House Effect or Global Warming

The gaseous mantle around our globe allows a considerable portion of solar radiations to enter right upto the surface of Earth which absorbs it and radiates back infra-red and heat waves. This heat is transferred to layers above, as warm layers rise and in turn passed onto higher and higher layers. Finally, much of the solar radiations are radiated back to space as infra-red and heat waves.

The system consisting of our globe and its atmosphere is in a state of dynamic equilibrium with the rate of absorption of solar radiations and its emission back to space as infra-red and heat waves, nearly balancing each other. Those gases and vapours which allow free passage to radiations of relatively shorter wavelengths (2900 Å–7000 Å) while absorbing effectively infra-red and heat waves (700 Å onwards) play a very important role in maintaining surface temperatures within a range in which life can exist. They form a blanket around a globe which checks the passage of infra-red and heat waves from earth crust back to space and keep it warm and hospitable. The phenomenon is similar to that of green house in which the glass enclosed atmosphere gets heated up due to its insulation from the rest of the environment. Hence, global warming is also known as **'Green House Effect'** and the gases responsible for it are called green house gases.

(A) Causes of Global Warming

Very slow, almost imperceptible rise of about 4–5°C in the global temperatures has occurred in the past 20,000 years. However, a rise of about 0.3–0.7°C was recorded during the last century alone, which is remarkably faster as compared to change that occured in the past.

(B) Gases Responsible for Global Warming

There are a number of gases present in the atmosphere which are capable of absorbing effectively heat waves and infra-red rays while being transparent to radiations of lower wavelengths. CO_2, CH_4, NO, NO_2, SO_2, O_3, CFC and water vapours are some of the gaseous constituents of troposphere which come in this category. From the point of global warming, however only those gases are important which maintain an effective concentration in the troposphere, *i.e.*, the region of atmosphere immediately covering Earth's surface. There are five such gases rising concentration of which has been implicated in causing noticeable rise in the mean global temperatures. These gases are CO_2, CH_4, NO and NO_2, chlorofluorocarbons (CFC) and water vapours. Other gases like SO_2, O_3 are not able to contribute much as they are quickly cleared from the atmosphere.

(*i*) *Carbon Dioxide:* It is one of the most important green house gases of which about 18 billion tons are being introduced into the atmosphere annually. A rise of about 26% has already been recorded in a period of 200 years only. Between the year 1980–90. CO_2 has been estimated to be responsible for atleast about 55% of global rise in temperature. The concentration of this gas is still rising at a rate of about 0.5% per year, as shown in Fig. 3.4(*b*).

(*ii*) *Methane:* Methane is another green house gas which is produced when organic matter decays under anaerobic conditions. In 1950 its concentration was about 1.1 ppm while in the year 1985, 1.7 ppm. The concentration of this gas is rising at a rate of about 1% per year. Between the year 1980–90 about 15% of the total warming has been attributed to this gas alone, as shown in Fig. 3.4(*d*).

In the atmosphere CH_4 undergoes oxidation to CO_2 and water both of which tend to accentuate the green house effect.

(*iii*) *Nitrous oxide:* It is another troublesome gas. Other oxides of nitrogen are reacted upon or cleared rapidly while nitrous oxides undergo decomposition very slowly and hence it tends to accumulate in the atmosphere. In 1950 its concentration was about 280 ppb while in 1985 it had reached 380 ppb. Between the years 1980–90 nitrous oxide accumulated for atleast 6% of the total global warming. The concentration of this gas is still rising at a rate of about 0.3% per year, as shown in Fig. 3.4(*e*).

(*iv*) *Chlorofluorocarbons:* CFC represent a group of man-made, colourless, odourless, easily liquefiable chemicals which have more potential for global warming than any other green house molecules. They are very stable compounds which may persist in the atmosphere for periods as long as 80–100 years. CFCs were responsible for 24% of global warming. Till 1985, about 15 million tons of these compounds had been released in the atmosphere. In spite of much international efforts to check the use of these chemicals CFCs are still rising at a rate of about 5% per year, as shown in Fig. 3.4(*c*).

(*v*) *Water vapours:* About 70% of Earth's surface is covered with water wherefrom an enormous quantity of water evaporates. Water vapours like any other green house gases contributes significantly to the global warming with an overall rise in temperatures the rate of global transpiration and evaporation.

This acceleration in the pace of global warming coincides with a rise in the concentration of green house gases in the atmosphere. The insulation of Earth's surface from the outer space caused by green house gases tends to become more and more effective as the concentration of these gases rises. More heat and infra-red radiations are trapped by the gaseous mantle around the globe which accelerates the pace of global warming, as shown in Fig. 3.4(a).

It is human activities which are mainly responsible for the accumulation of green house gases in the atmosphere. Energy production and its use, intensive agriculture, maintenance of huge livestock population, use of CFCs, land use modification and industrial production are some of the aspects of human activity which are responsible for this accumulation.

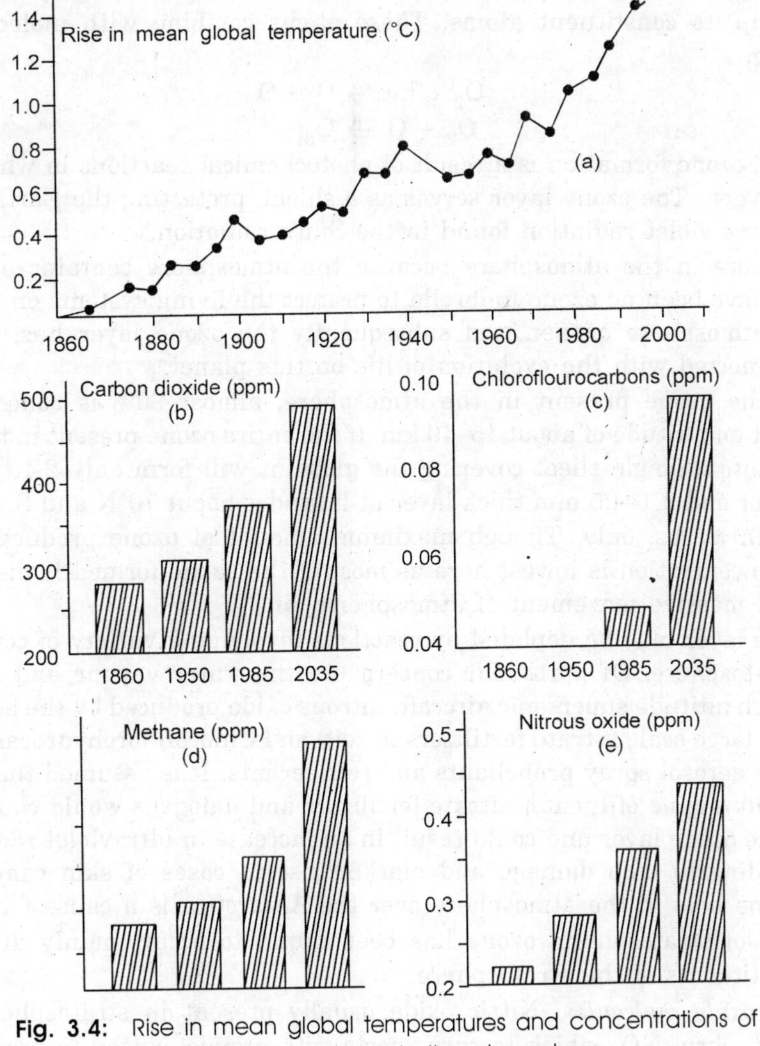

Fig. 3.4: Rise in mean global temperatures and concentrations of Green House gases in the atmosphere

(C) Consequences of the Green House Effect

Following are the main results of the green house effect:

(*i*) Temperature changes

(*ii*) Melting of polar ice cap

(*iii*) Rising of sea levels

(*iv*) Climatic changes

(v) Crop production and nutrient depletion

3.1.10 Formation and Depletion of Ozone Layer

High up in the stratosphere, about 15–40 km above Earth's surface, short wavelength ultraviolet radiations in the range of 1800 Å to 2200 Å are absorbed by molecular oxygen which splits up its constituent atoms. These atoms combine with molecular oxygen to produce Ozone:

$$O_2 + h\nu \rightarrow O + O$$
$$O_2 + O \rightarrow O_3$$

Therefore, ozone formation is a result of photochemical reactions in which the starting molecule is oxygen. The ozone layer serves as a shield, protecting the Earth's surface from most of the ultra violet radiation found in the Sun's radiation.

Ozone occurs in the atmosphere because the atmosphere contains oxygen. Without oxygen there have been no ozone umbrella to protect the living systems on Earth's surface. Evolution of atmosphere oxygen and subsequently the ozone layer has, therefore, been intimately connected with the evolution of life on this planet.

Most of the ozone present in the atmosphere, almost 90% is concentrated in the stratosphere at an altitude of about 15–40 km. If the entire ozone present in the stratosphere is condensed into a single sheet covering the globe, it will form only 2.4 to 2.6 mm thick layer at equator and 3.1–4.5 mm thick layer at latitudes about 70°N and S. At ground level ozone occurs in traces, only. Though maximum amount of ozone production occurs over equator, its concentration is lowest here as most of the ozone formed is displaced towards the poles with massive movement of atmospheric air.

The ozone layer may be depleted by reactions involving a variety of compounds which reach the stratosphere. Of particular concern are the water vapour and nitrogen oxides released by high altitude supersonic aircraft, nitrous oxide produced by the action of bacteria in soils (use of large scale nitrate fertilizers in soil) and Chlorofluorohydrocarbons which are widely used as aerosol spray propellants and refrigerants. It is assumed that the combined effect of aircraft engine effluents, nitrate fertilizers and halogens would cause considerable depletion of the ozone layer and could result in an increase in ultraviolet radiation reaching the Earth leading to crop damage and marked rise in cases of skin cancer. The recent reports of ozone hole in the atmosphere over the Antarctica is a cause for great concern.

Depletion of stratospheric ozone has been found to occur mainly due to the three following constituents of the stratosphere:

(i) *Nitric oxide molecules:* Nitric oxide usually present in stratosphere, reacts with ozone to form NO_2 which in turn reacts with atomic oxygen to produce NO again.

$$NO + O_3 \rightarrow NO_2 + O$$
$$NO_2 + O \rightarrow NO + O_2$$

(ii) *Chlorine atoms:* Chlorine atoms react with ozone to yield chlorine monoxide (ClO) which reacts with atomic oxygen to regenerate chlorine atoms again.

$$Cl + O_3 \rightarrow ClO + O_2$$
$$ClO + O \rightarrow Cl + O_2$$

(iii) *Hydroxyl ions:* Hydroxyl ions produced by photodissociation of water molecules in the stratosphere, reacts with ozone molecules to produce HO_2 which reacts with another O_3 molecule to yield hydroxyl ions again.

$$OH + O_3 \rightarrow HO_2 + O_2$$
$$HO_2 + O_3 \rightarrow OH + 2O_2$$

The above reactions are not so simple. The chemical processes which occur under conditions of low temperature, low pressures and plenty of high energy UV radiations are exceedingly complex in nature and involve almost a hundred and fifty independent reactions.

(A) Causes of Stratospheric Ozone Depletion

The three main reactants, namely, nitric oxide, chlorine atoms and hydroxy ions are normally present in the environment. Though much of these is derived from natural sources, human activity is also equally responsible for contributing a lot of such constituents in the atmosphere which give rise to these reactions. Temperature decreases with height in the troposphere. In between the troposphere and stratosphere, lies a narrow region, called the tropopause, which has almost uniform temperature. The tropopause layer (region) prevents the exchange of materials between troposphere and stratosphere. However, some exchange does occur and many constituents of troposphere do reach the stratosphere. In addition to this exchange, many pollutants are injected straight into the stratosphere by supersonic transport planes, rockets and space shuttles, surface nuclear tests, volcanic emissions etc. Human activities responsible for polluting the stratosphere may be summed up as follows:

(i) Combustion of Fossil Fuels and Organic Matter: Huge quantities of various oxides such as oxides of carbon, nitrogen and sulphur, hydrocarbons and fine particulate material is introduced daily into the troposphere some of which goes high enough to reach the stratosphere. To this may be added microbial decomposition products of organic matter which discharge plenty of hydrocarbons into the atmosphere of which CH_4 is one of the most important constituents which reaches high enough to pollute the stratosphere.

(ii) Excessive use of Nitrogenous Fertilizers: Microbial action on nitrogenous fertilizers produces nitrous oxide which escapes into the atmosphere. As this gas is decomposed with difficulty it tends to accumulate in the atmosphere. The estimated tropospheric load of nitrous oxide is 1.7×10^{18} gm while the stratospheric content is about 8.0×10^{15} gm. Action of high energy radiations converts it to nitric oxide which is an important ozone destroying constituent of the stratosphere.

(iii) Excessive use of Chlorofluorocarbons: Chlorofluorocarbons are inert, highly stable, colourless, odourless chemicals which can be easily liquefied. These are mostly halogenated hydrocarbons which under the influence of high energy radiations break up to yield a chlorine atom. A single chlorine atom can destroy 1,00,000 ozone molecules.

Some important chlorofluorocarbons which are responsible for ozone depletion are given in the Table 3.9.

Today, about 16 billion kg of CFCs have been produced and used up; these are seriously depleting the Earth's protective shield. Earlier models predicted a 6.5% ozone depletion by the year 2030; but it could well be as much as 16%. A rule of thumb equates a 1% depletion in ozone with a 2% increase in UV rays and 4% increase in the potential for skin cancer.

Table 3.9: Ozone depleting chemicals (CFCs)

Compounds	*Chemical Composition*	*Chemical Names*
Man-made Sources:		
CFC–11	$CFCl_3$	Trichloro fluoromethane
CFC–12	CF_2Cl_2	Dichloro difluoromethane
CFC–13	$C_2F_3Cl_3$	Trichloro trifluoroethane
—	CCl_4	Carbon tetrachloride
—	CH_3Cl_3	Methyl chloroform
CFC–22	$CHClF_2$	Difluoro Chloromethane
Halon–1211	CF_2ClBr	Difluorochloro bromomethane
Halon–1301	$CBrF_3$	Trifluoro bromomethane
Natural Sources:		
—	CH_3Cl	Methylchloride

(*iv*) *Supersonic Transports, Rockets and Space Shuttles:* Supersonic jet-liners, which fly in lower regions of the stratosphere discharge various oxides of carbon, nitrogen, sulphur, hydrocarbons and particulate material. Space flights which are powered by powerful rockets also release large quantities of these pollutants right into the atmosphere. Ammonium percolate used in rockets as oxidant releases plenty of chlorides.

(*v*) *Nuclear Tests:* Surface nuclear explosions produce huge quantities of various gases, dust shoot and debris with enormous force which carries much of material straight into the stratosphere. Much of these materials damage the ozone layer.

The three reactants which play a major role in destruction of ozone layer, namely, nitric oxide, chlorine atoms and hydroxyl ions may also be derived from natural sources. Major natural sources of pollutants which are responsible for stratospheric zone depletion are solar flares and volcanic eruption.

(B) Consequences of Stratospheric Ozone Depletion

The following would be the consequences of ozone depletion:

(*i*) Eye damages including tumours of the conjuctive and cornea, cataracts and retinal damage may be enhanced.

(*ii*) An 1% boost in UV radiation due to losses could lead to 5% increase in the skin cancer rates. Moreover, a number of skin diseases are either triggered by UV exposure or are manifested by excessive sensitivity to UV radiation.

(*iii*) Important crops (wheat, corn, rice and soyabeans) are also sensitive to UV rays. A rule of thumb estimates a 1% reduction in crop yields for every percent or ozone depletion.

(*iv*) Ozone depletion could harm aquatic organisms, exacerbate smog in cities, and accelerate the degradation of some plastics and paints.

(*v*) The ozone depletion chemicals (CFCs) and halons) have a 10,000 times more heat-absorption power than oxygen. At current emission rates, the heating potential of CFCs will rise from 15% to the equivalent of CO_2 by the year 2020.

(*vi*) Photochemical changes occurring in such biological molecules as nucleic acids, proteins, lipids, steroids and melanins may cause damage to the cell nucleus, cytoplasm, organelles and membranes.

(*vii*) Inflammatory changes, numerous biochemical interactions, involving reactive oxygen species and immunologic changes including the supression of certain immune responses and various repair processes may follow.

3.1.11 Radiation Balance of the Earth

The Earth receives a very large input of energy daily from the Sun and maintains a steady state by giving off the bulk of this energy at the same rate. The solar spectrum incident on the Earth's surface has a peak at 483 cm, *i.e.*, visible region, indicating the surface temperature of the Sun as approx. 6000°K. The Earth's spectrum shows a peak of about 10,000 nm *i.e.*, infra-red region, with the surface temperature of the Earth about 29°K (17°C). In other words, the Earth absorbs radiation mainly in the visible region but emits radiation at the same rate in the infra-red region.

The solar flux incident on the Earth's Supper atmosphere is 1340 Watts m^{-2} min^{-1}. The Earth absorbs about 66% of the solar flux incident on it, *i.e.*, 19.5 kCal m^{-2} min^{-1} (or 81.6 kJ m^{-2} min^{-1}), while it reflects and scatters back into space about 34% of the solar flux. Energy transport plays a crucial role in the Earth's radiation balance. This proceeds through the mechanism of:

(*i*) Radiation of energy in the IR region from the Earth.

(*ii*) Conduction of energy through the interaction of atoms or molecules, and

(*iii*) Convection of energy through massive air circulation.

The latter two mechanisms are responsible for loss of heat from the Earth's surface through transport to clouds, and consequent radiation from the clouds.

The average surface temperature of the Earth is maintained around 15°C due to reabsorption of most of the outgoing infrared radiation (2–40 μ) by water vapour (4–8 μ), and to a lesser extent by CO_2 (12–16.3 μ). The combined effect of water vapour and CO_2 leads to the green house effect, which is of great significance in governing the climate on Earth.

Increasing agricultural and industrial outputs can also upset the Earth's radiation balance by changing the albedo (fraction of Sun light reflected and scattered back to the atmosphere). Deforestation and the consequent soil erosion increase the albedo.

3.1.12 To Lapse Rate and Temperature Inversion

(A) Lapse Rate

The change of temperature with height in the troposphere is called the lapse rate. In the troposphere region, the temperature decreases fairly steadily from the ground temperature to a temperature of –50 to –80°C at a rate of about 6.5°C/km, which is referred to as the lapse rate of the troposphere, as shown in Figure. 3.5.

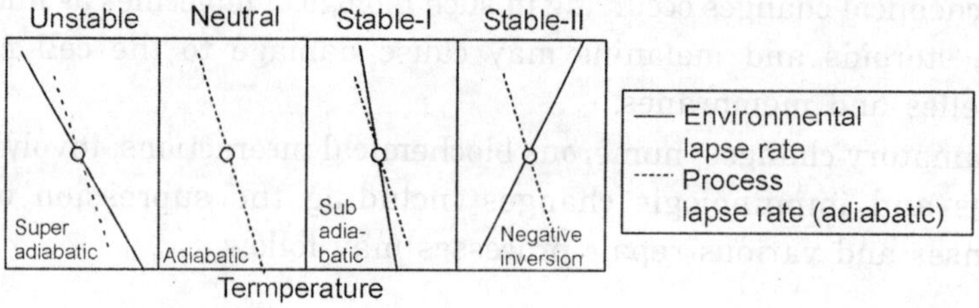

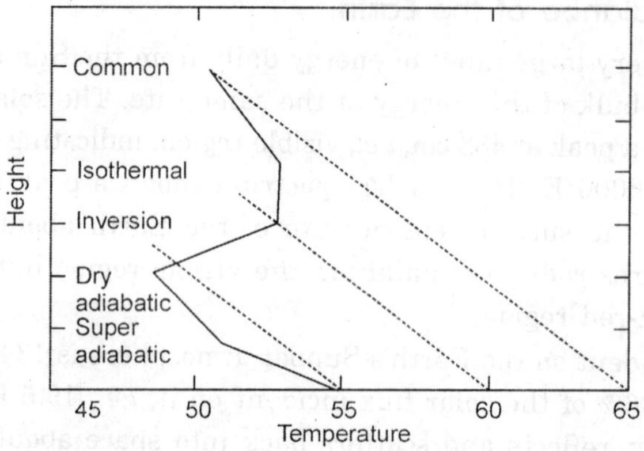

Fig. 3.5: Lapse rates on temperature profile

As we go higher in the atmosphere, the height or weight of air above decreases and so does the pressure. Any gas, free to expand, will tend to move from the region of high pressure to the region of low pressure. As it does so, it expands and, if there is no interchange of heat with the surroundings, the gas will become cooler. This is called adiabatic cooling. This adiabatic lapse rate can be obtained by considering the pressure-volume-temperature relationship, the adiabatic expansion and the change of pressure with height.

Equation of state $PV = nRT$

Change of pressure with height $dp/dh = -g\dfrac{nM}{V}$

Law of adiabatic expansion $P' =$ constant

or $T/T_o = (P/P_o)^{R/Cp}$

Now, adiabatic lapse rate $= dT/dh = dT/dp \times dp/c'' = -g\dfrac{nM}{V}\left[\dfrac{dT}{dp}\right]$

Where dT/dp can be obtained as

$$\frac{dT}{dp} = T_o \frac{R}{C_p} \times \frac{P^{R/C_p - 1}}{P_o^{R/C_p}} = \frac{T_o T}{T} \cdot \frac{R}{C} \cdot \frac{P}{P_o^{R/C_p - P}}$$

$$= \frac{R}{C_p} \cdot \frac{T}{P}$$

Now, substituting the value of dT/dp and dp/dh, we get

$$\frac{dT}{dh} = \frac{R}{C_p} \cdot \frac{T}{P}\left(-g\frac{nM}{V}\right) = -g\frac{M}{C_p}$$

as $$n\frac{RT}{PV} = 1$$

and the values of R = 8.314 J/mole °K

$C_p = 3.5$

R = 3.5 × 8.314 J/mole °K

M = 28.9 and g = 9.8 m/s^2

On substitution, this value works out to be –9.8°C/km for dry air. If the air is moist, the water present may undergo a phase change from vapour to liquid. When a moist parcel of air rises, expands and cools, the water vapour can condense and release latent heat. Thus, the actual moist adiabatic lapse rate is less than the dry rate and depends on humidity. It is the range of –6 to –8°C/km with higher values occurring at greater heights where the water content is less, as shown in Figure. 3.6.

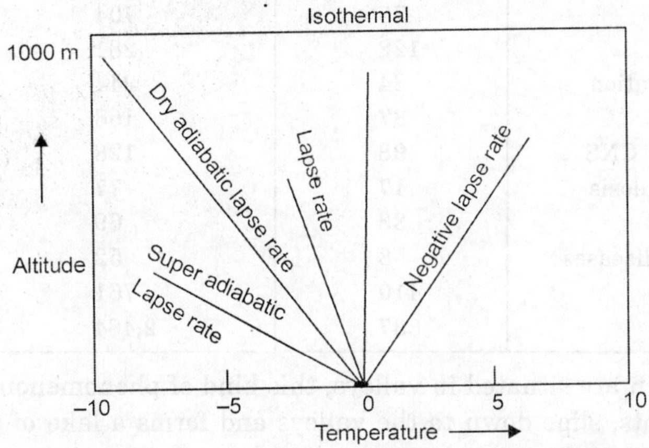

Fig. 3.6: Position of lapse rate under different atmospheric conditions

On a clear summer day, rapid heating of the Earth by the Sun warms the air near the surface to the point where the lapse rate is super adiabatic. Under these conditions, the atmosphere is said to be in unstable equilibrium and marked vertical mixing of air results. This is the condition in which pollutants are rapidly dispersed.

(B) Temperature Inversion

The opposite condition of the lapse rate is the negative lapse rate of temperature inversion. In this case, the temperature increases as we go higher. This condition is most undesirable for the dispersion of pollutants.

Inversion takes place in the atmosphere due to two reasons:

(*i*) Radiation inversion
(*ii*) Subsidence inversion

(*i*) *Radiation Inversion*

The air near the surface cools during night by long wave radiation. Thus, the temperature is lower near the ground than at a height. This is called radiation inversion. When the Sun rises and heats air near the ground, the inversion disappears, as shown in Fig. 3.6.

(ii) *Subsidence Inversion*

This kind of inversion results when the upper layer of air descends (subsides) during a developing anticyclone (high pressure centres). The subsidising air warms as it contracts but the warming will be greater at the upper part than at the lower part near the ground.

Many of the episodes of air pollution which have taken place in the world are due to inversion. Details of the famous London Smog of 1952 that killed 4000 peoples and lasted for 5 days are given in Table 3.10.

Table 3.10: Mortality increases in the London smog of 1952

Cause	*Seasonal Norm*	*No. of deaths in the week after Smog*	*Relative Increase*
Bronchitis	75	704	9.4
Coronary disease	122	281	2.3
Myocardial degeneration	84	244	2.9
Pneumonia	37	168	4.5
Vascular lesions of CNS	98	128	1.3
Respiratory tuberculosis	17	77	4.5
Cancer of lung	36	69	1.9
Other respiratory diseases	8	52	6.5
Other causes	410	761	1.9
All causes	887	2,484	2.8

In towns, which are situated in valleys, this kind of phenomenon, where the dense air, filled with pollutants, slips down to the valleys and forms a lake of polluted air, is a very harmful situation.

3.1.13 Pollutants

Pollutants are chemical, physical or biological agents that exert undesirable effects on human health, his environment or belongings. The air we breathe, the water we drink, the food we consume and the land we live on are all contaminated with one form or other of pollutants, as shown in Table 3.11.

Table 3.11 : Types of environmental pollutants

1. Chemical agents	Gases and particulates Heavy metals Pesticides Petroleum Solid and liquid wastes
2. Physical agents	Heat Odour Noise Radiation
3. Biological agents	Microorganisms Population

Several chemicals which are now regarded as pollutants by man have been present on the Earth as part of the natural core. Heavy metals and radioactive elements are some of the naturally occurring substances. Only when their levels entering the environment increase, (which normally is accompanied by the accelerated anthropogenic inputs), they are considered as pollutants, as shown in Fig. 3.7.

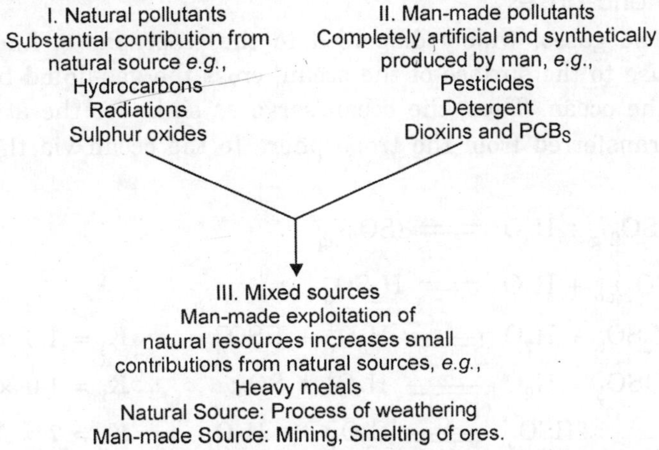

Fig. 3.7: Natural and man-made sources of pollution

Therefore, a pollutant has been defined as: "*A substance or effect is normally considered to be a pollutant if it adversely alters the environment by changing the growth rate of a species, interferes with the food chain, is toxic, or interferes with health, comfort, amenities or property values of people.*"

Generally, a pollutant is a substance or effect introduced into the environment in significant amounts as sewage, waste, accidental discharge or as a by product of manufacturing process or other human activity. A polluting substance can be a solid, semi-solid, liquid, gas or sub-molecular particle. A polluting effect is normally some kind of waste energy such as heat, noise or vibration.

Contaminants

A material which does not occur in nature, but is introduced by human activities into the environment, affecting its composition. A contaminant is classified as a pollutant when it exerts a detrimental effect, *e.g.*, chlorine gas escaped from a derailed railway tank car near Youngstown. Florida in 1978 and killed eight persons driving on a nearby road. This gas does not occur in the atmosphere, so it is undoubtedly a contaminant because of its dangerous effect; it is also a pollutant.

Receptor

It is a medium which is affected by a pollutant. Man is the receptor of photochemical smog causing irritation of the eyes and respiratory tract. Trout fingerlines are receptors of dieldrin in water which causes their death.

Sink

The medium which retains and interacts with a long-lived pollutant. A marbel wall will act as a sink for atmospheric sulphuric acid and ultimately get damaged.

$$H_2SO_4 + CaCO_3 \rightarrow CaSO_4 + H_2O + CO_2$$

Sinks of atmospheric gases

(*a*) *Oceans:* The gases, widely dispersed in the troposphere, find ready sink in the oceans. They diffuse to the surface of the ocean, cross the gas-liquid boundary and finally into the bulk of the ocean. Thus, the ocean serve as sinks for the atmospheric gases.

The SO_2 is transferred from the troposphere to the ocean via the following reaction route:

$$(SO_2)_g + H_2O \rightleftharpoons (SO_2)_{aq}$$

$$(SO_2)_{aq} + H_2O \rightleftharpoons H_2SO_3$$

$$H_2SO_3 + H_2O \overset{K_1}{\rightleftharpoons} H_3O^+ + HSO_3^-; \quad K_1 = 1.6 \times 10^{-2} \text{ mole}^{-1}$$

$$HSO_3^- + H_2O \overset{K_2}{\rightleftharpoons} H_3O^+ + SO_3^-; \quad K_2 = 1.0 \times 10^{-7} \text{ mole}^{-1}$$

$$2HSO_3^- \overset{K_3}{\rightleftharpoons} S_2O_5^{--} + H_2O; \quad K_3 = 7 \times 10^{-2} \text{ mole}^{-1}$$

Recent data estimate the annual transfer of SO_2 to the oceans as 4×10^7 tons per year.

(*b*) *Vegetation:* Plants have the ability to take up atmospheric gases without active metabolism. Some plants may actively metabolize the gas thus creating an optimum concentration gradient for further absorption. Extensive laboratory experiments have been undertaken to determine the velocity of deposition, V_g:

$$V_g = \frac{\text{Total deposition per unit area}}{\text{Dosage (Concentration} \times \text{time)}}$$

Maize, grassland, bean and alfalfa show uptake of O_3 (0.2 g/m^2). SO_2 (0.2 g/m^2), CO (0.52 g/m^2) and NO_2 (0.17 g/m^2) respectively. The corresponding values of V_g (cm/sec) are 1.25, 0.8, 0.012 and 2.0.

3.2 WATER POLLUTION

Water is essential for the survival of any form of life. A large quantity of water, used for domestic and industrial applications, require certain specifications. Chemical and allied industry requires a large amount of water for their condensors and coolers. The engineering industry also requires a considerable amount of water for different applications including the generation of steam. In power houses too, water is used for generation of high pressure steam. Besides the above, water is used for recreational purposes.

The large quantity of water is drawn from river, lakes etc. which is contaminated due to the effluents received from industry. Many of the effluents thus joining the river are biologically degraded by the flora and fauna present in the river. It is estimated that out of total rainfall received annually by India, 96% water used for agriculture, 3% for domestic use and 1% for industry activity. An analysis conducted in 1982 revealed that about 70% of all available water in India is polluted.

3.2.1 Sources and Cause of Water Pollution

Polluting substances in natural sources of water have their origins in a great variety of places. These sources of contamination of water are classified into following categories; Natural pollution, Agricultural pollution, Mining pollution, Municipal pollution and industrial pollution.

Natural Pollution

This type of pollution is either accidental or occasional. Aerial contaminants, like gas and dust added to water in the form of rainfall, soil silt stripped from the land by heavy rainfall or melting of snow or as soluble salts leached from the earth by surface streams. Leachates from animal excreta, decaying bodies of plants and animals, deep ponds etc. also introduces micro-organisms to natural water.

Agricultural Pollution

The use of fertilizers, insecticides, herbicides and weed killers in agricultural practices, poses serious pollution problems as most of them are resistant to natural degradation.

Mining Pollution

Mines or tailing from ore washing are disposed off in water suspensions, which ultimately transferred to the natural stream which thus gets polluted. Sometimes mining operations produces soluble toxic materials.

Municipal Pollution

The pathogenic bacteria, suspended solids and oxygen consuming organic matter are transferred to natural stream through municipal wastes in the form of domestic sewage includes the discharge from toilets, wash room, kitchens, institutional, commercial and industrial building wastes, causes water pollution.

Industrial Pollution

The manufacturing plants in industry sector utilised large quantity of water. Out of which only a small part of it may appear in the final product. Rest unused waste. Water containing waste as well as toxic materials comes into the main stream of natural water causing severe pollution.

3.2.2 Types of Water Pollutants

Water pollutants may be usually classified into following five categories:

(*i*) Organic Pollutants
(*ii*) Inorganic Pollutants
(*iii*) Suspended Solids and Sediments

(*iv*) Radio active Materials

(*v*) Thermal or Heat Pollutants

3.2.2.1 Organic Pollutants

Organic pollutants may be further classified into following five sub categories:

Synthetic organic compounds

These are the man-made materials such as synthetic pesticides, pharmaceuticals, insecticides, paints, detergents solvents food additives, plastics, elastomers, fibres etc. Most of these chemicals are potentially toxic to plants, animals and humans. Aromatic chlorinated hydrocarbons, non degradable chemicals such as alkyl benzene sulphonate from synthetic detergent, volatile substances such as alcohols, aldehydes, ethers, gasoline may cause explosion, colour, odour in sewers.

Organic-demanding wastes

It includes the domestic and animal sewage, bio-degradable organic compounds and industrial wastes from food processing plants, meat packing plants, slaughter houses, pulp and paper miles, tanneries etc. as well as agricultural wastes. The above mentioned wastes undergoes degradation and decomposition by bacterial activity in the presence of dissolved oxygen *i.e.*, DO, resulting depletion of DO which is harmful to aquatic organisms.

Disease causing wastes

Such type of wastes include pathogenic microorganisms which may enter into the water stream alongwith wastes and sewage causing damage to public health. These microbes mainly viruses and bacteria causes diseases such as cholera, typhoid, dysentery, polio, hepatitis etc.

Oils pollutants

Oil pollution may takes place either accidently or occasionally through oil tankers, oil spills, fires in ship and oil tankers, leakage from oil pipelines etc. The oil wastes added to natural water stream may causes reduction in DO which endangers water birds, coastal plants and animals.

Sewage and agricultural run-off

Sewage and agricultural run-off also reduces the DO level of water in the long run due to the natural biological process of eutrophication and ends up as a dead pool of water.

3.2.3 Effects of Water Pollutants on Living Beings

The effects of water pollution on public health and safety, effects on aquatic and other life, damage to property and other economic losses are described below in brief.

3.2.3.1 *Effect on Public Health and Safety*

The principal hazards to public health from polluted water are mainly due to presence of pathogenic bacteria from domestic sewage, toxic materials such as acids, alkalis, cyanide and heavy metals as well as materials from industrial wastes. Danger to public health from polluted water comes not only from direct use of water but also from consuming food like fish that have lived in polluted water.

3.2.3.2 *Effect on Aquatic and Other Life*

Pollution of natural water may be disastrous to fish and other organisms naturally inhabiting the water stream. Organic wastes under the influence of bacterial action deplete the dissolved oxygen by consuming it in biochemical oxidation reaction, resulting in the cause of danger to fish and other aquatic animals. Suspended solids such as silt and coal have been particularly troublesome in this respect.

3.2.3.3 *Damage to Property*

Industrial waste and discharge of acids leads to damage of property through corrosive attack, e.g., ships, building, and other concrete structure, water piping, pumps, boilers, coolers, aquaguard etc.

3.2.3.4 *Other Economic Losses*

Polluted streams which are used as the only source of municipal or industrial water require substantial treatment at considerable cost for the removal of suspended solids, taste, odour, colour, hardness and specific chemicals present.

3.2.4 Sewage Treatment

The dissolved organic matter and its consequences for dissolved oxygen in the aquatic environment, which receives the effluent, many other materials like bacteria, viruses and dissolved inorganic salts may also be present which can cause problems in the environment. Methods are known which can completely purify waste water, even sewage to such an extent that it can be safe for drinking. Such procedures are only applicable for large cities with limited supply of water, because the cost of the installation is of course extremely high.

The waste water treatment involves the following procedures:

(*i*) Primary treatment

(*ii*) Secondary treatment (settling process) and

(*iii*) Tertiary treatment

Primary and secondary treatment process including activated sludge treatment and sludge digestion are also very expensive. This involves the removal of nearly all particles and organic matter present but excludes. The dissolved salts, many of which are plant nutrients. Bacteria however tend to survive this treatment in enough numbers to constitute a possible health hazards in the effluent. The same applies to viruses, in particular the hepatitis virus. The high temperature anaerobic sludge digestion process which can yield methane as a useful product is a desirable addition to a waste water treatment plant. It tends to kill the disease causing organisms.

The complete treatment of waste water is usually divided into three stages *viz.* primary, secondary and tertiary treatments. The final level of purity depends on the economics of the situation. It would naturally cost more to refine the water after the second and the third stages so as to make the water fit for drinking.

3.2.4.1 *Primary Treatment*

Waste water usually contains large quantities of floating rubbish such as cloth, wood and cans. These larger objects are removed by metal bars which act like stainers as the water moves beneath them in an open channel. The velocity of the water is then reduced in a grit settling chamber of a larger size than the previous channel. This results in most of the grit and sand settling on the bottom. Periodically this settling chamber is disconnected from the main system and the grit is removed manually.

The smaller and lighter particles take longer time to settle under gravity. This slower process takes place in the primary sedimentation tanks. Such tanks are quite large and are usually circular with a bottom shaped like a shallow cone.

At the bottom of the cone, the sediments or sludge is drained away by a horizontal pipe. Rotating scrapers slowly scrape the sludge from the sloping sides, moving it gradually from the perimeter to the centre where it is drained off as described above. The disposal of this primary sludge can be a problem because of its offensive smell. It also has a water content of over 95%.

It must be dried before it can be disposed of on land. Alternatively it may be digested by the method described below under the heading "sludge digestion". The liquid effluent from the primary sedimentation tanks has now lost over half of its original suspended matter. This effluent may be drained directly into a river or the sea if other considerations permit this.

3.2.4.2 *Secondary Treatment*

The suspended particles remaining after primary treatment are too fine to be removed by a simple sedimentation process. This would take far too long and much of the materials is colloidal and would never settle down. Most of this material is, however, organic and this means that certain micro organisms can consume it as food. The carbon chains of the large organic molecules are thus broken down to smaller molecules such as CO_2. This can be accomplished in a trickling filter or in an activated sludge digestion.

Incoming waste water (after primary treatment) is sprayed over circular filter beds by a rotating boom. This is a pipe with holes along its length and the water emerges in horizontal jets. The reaction to these jets causes the boom to rotate thus no additional power is required. The water is kept at a required pressure head by a type of siphon at the inlet.

The filter bed is usually about 1.5 m deep and surrounded by a circular brick wall. The filter consists of various combinations of stony material or even plastic material. Stones are irregular in shape, possibly round and of 3 to 10 cm diameter.

One of the problems which could be faced if the organisms which grow on the stones eventually block the filter if too much effluent of high BOD is applied.

It is important to note that trickling filters, like other forms of secondary treatment, require aercbic conditions. If there is a lack of oxygen, the wrong types of organisms would develop and toxic materials would find their way into the final effluent. Under ideal conditions the filter bed should provide a home for a wide variety of organisms including bacteria, protozoa, fungi, worms and insects. The upper surface will also be colonized by algae.

Activated sludge process

The output of primary treatment is usually treated by the activated sludge process if large volume of effluent have to be treated. The object here is to provide oxygen to a large volume of effluent after primary treatment. The aeration can be carried out by forced air or mechanical stirrers. The water is kept in constant turbulent motion in concrete tanks which are 5 m deep and 10 m wide. The length is 50 to 100 m.

The result of this process is to suspend an enormous number of sludge particles and organic material in an aerobic environment. Each particle then becomes colonised by aerobic micro organisms.

It is important to note that the incoming water does not contain sludge. Therefore, the tanks must be constantly fed with sludge to provide the suspended particle without which the process would be much slower.

Sludge from the final settling tank is recycled to the extent of about 20% of the incoming volume to the aeration tank. Surplus sludge will eventually accumulate in the final settling tank and this has to be disposed off, or can sometimes be used as fertilizer after drying.

Sludge digestion

An attractive alternative to sludge disposal is the digestion of sludge in the absence of oxygen at 35°C. Under these conditions (known as mesophilic) the micro organism converts the organic matter including the methane gas into simpler compounds. About 2/3rd of the sludge in one batch is consumed in this way and the process is completed in about three weeks. The methance (CH_4) formed can be used as fuel to provide the heat on burning, required to warm the digestion tanks.

The sludge which remains after digestion is inoffensive.

3.2.4.3 *Treatiary Treatment*

Tertiary treatment is used to remove the dissolved materials which have survived the primary and secondary treatment. The most important solutes used in this process are compounds of phosphorous and nitrogen. Phosphorous is more important than nitrogen in causing eutrophication, which is the process results in excessive and undesirable growth of plant life in lakes and harbours. Other toxic substances may also be present and special processes are used in each case.

The end result should be water which is suitable for drinking.

Phosphorous can be removed by precipitation with coagulants such as $Al_2(SO_4)_3$ and $FeCl_3$. This is sometimes carried out in the activated sludge step of the secondary treatment, but in true tertiary treatment, this precipitation is a separate process. Lime as $Ca(OH)_2$, can also be used to precipitate phosphorous.

Nitrogen is more difficult to remove than phosphorous, but were it is in the form of NH_3 it can be stripped by air at pH 11.0. This means that effluent from the lime treatment is particularly suitable for ammonia removal.

A part from nitrogen and phosphorous, there are a variety of substances which contribute to the appearance and even the taste of the water. These can be removed by filtration through certain media, including activation charcoal. The charcoal can be reactivated and reused.

Many of the substances which require removal by tertiary treatment can theoretically be removed by biological process, particularly in conjunction with the activated sludge step of secondary treatment.

3.3 GREEN CHEMISTRY

Green chemistry is a chemical philosophy encouraging the design of products and processes that reduce or eliminate the use and generation of hazardous substances. Whereas environmental chemistry is the chemistry of the natural environment, and of pollutant chemicals in nature, green chemistry seeks to reduce and prevent pollution at its source. In 1990 the Pollution Prevention Act was passed in the United States. This act helped to create a *modus operandi* for dealing with pollution in an original and innovative way.

As a chemical philosophy, green chemistry derives from organic chemistry, inorganic chemistry, biochemistry, analytical chemistry, even physical chemistry. However, the philosophy of green chemistry tends to focus on industrial applications. Contrast this with click chemistry which tends to favor academic applications, although industrial applications are possible. The focus is on minimizing the hazard and maximizing the efficiency of any chemical choice. It is distinct from environmental chemistry which focuses on chemical phenomena in the environment.

In 2005 Ryoji Noyori identified three key developments in green chemistry: use of supercritical carbon dioxide as green solvent, aqueous hydrogen peroxide for clean oxidations and the use of hydrogen in asymmetric synthesis. Examples of applied green chemistry are supercritical water oxidation, on water reactions and dry media reactions.

Bioengineering is also seen as a promising technique for achieving green chemistry goals. A number of important process chemicals can be synthesized in engineered organisms, such as shikimate, a Tamiflu precursor which is fermented by Roche in bacteria.

Principles of Green Chemistry

Paul Anastas, then of the EPA, and John C. Warner developed 12 principles of green chemistry, which help to explain what the definition means in practice. The principles cover such concepts as:

- The design of processes to maximize the amount of raw material that ends up in the product;
- The use of safe, environment substances, including solvents, whenever possible;
- The design of energy efficient processes;
- The best form of waste disposal: do not to create it in the first place.

The 12 principles are:

1. *Prevent waste:* Design chemical synthesis to prevent waste, leaving no waste to treat or clean up.
2. *Design safer chemicals and products:* Design chemical products to be fully effective, yet have little or no toxicity.
3. *Design less hazardous chemical synthesis:* Design synthesis to use and generate substances with little or no toxicity to humans and the environment.

4. *Use renewable feedstock:* Use raw materials and feedstock that are renewable rather than depleting. Renewable feedstock are often made from agricultural products or are the wastes of other processes; depleting feedstock are made from fossil fuels (petroleum, natural gas, or coal) or are mined.
5. *Use catalysts, not stoichiometric reagents:* Minimize waste by using catalytic reactions. Catalysts are used in small amounts and can carry out a single reaction many times. They are preferable to stoichiometric reagents, which are used in excess and work only once.
6. *Avoid chemical derivatives:* Avoid using blocking or protecting groups or any temporary modifications if possible. Derivatives use additional reagents and generate waste.
7. *Maximize atom economy:* Design synthesis so that the final product contains the maximum proportion of the starting materials. There should be few, if any, wasted atoms.
8. *Use safer solvents and reaction conditions:* Avoid using solvents, separation agents, or other auxiliary chemicals. If these chemicals are necessary, use innocuous chemicals. If a solvent is necessary, water is a good medium as well as certain eco-friendly solvents that do not contribute to smog formation or destroy the ozone.
9. *Increase energy efficiency:* Run chemical reactions at ambient temperature and pressure whenever possible.
10. *Design chemicals and products to degrade after use:* Design chemical products to break down to innocuous substances after use so that they do not accumulate in the environment.
11. *Analyze in real time to prevent pollution:* Include in-process real-time monitoring and control during synthesis to minimize or eliminates the formation of byproducts.
12. *Minimize the potential for accidents:* Design chemicals and their forms (solid, liquid, or gas) to minimize the potential for chemical accidents including explosions, fires, and releases to the environment.

3.4 EUTROPHICATION

Eutrophication refers to an increase in the primary productivity of any ecosystem. It is caused by the increase of chemical nutrients, typically compounds containing nitrogen or phosphorous. It may occur on land or in water. Therefore, any factor that causes increased nutrient concentration can potentially lead to eutrophication.

Eutrophication is frequently a result of nutrient pollution such as the release of sewage effluent into natural waters although it may also occur naturally in situations where nutrients accumulate or where they flow into systems on an ephemeral basis. Eutrophication generally promotes excessive plant growth and decay, favours certain weedy species over others, and is likely to cause severe reductions in water quality. In aquatic environments, enhanced growth of choking aquatic vegetation or phytoplankton disrupts normal functioning of the ecosystem, causing a variety of problems such as lack of oxygen in the water for fish and shell-fish to survive. The water then becomes cloudy, coloured a shade of green, yellow brown, or red.

Eutrophication was recognized as a pollution problem in European and North American lakes and reservoirs in the mid 20th century. Since then, it has become more widespread.

Survey showed that 54% of lakes in Asia are eutrophic; in Europe, 53%; in North America, 48%, in South America, 41%, and in Africa, 28%.

Phosphorous is often regarded as the main culprit in cases of eutrophication in lakes subjected to point source pollution from sewage. The concentration of algae and the trophic state of lakes correspond well to phosphorous levels in water. It has been observed that between 1950 and 1955, 600,000,000 tonnes of phosphorous were applied to earth's surface, primarily on croplandes. Human kind has increased the rate of phosphorous cycling on earth by four times, mainly due to agricultural fertilizer production and application.

Like lakes, rivers, and oceans, terrestrial ecosystems are subject to similarly adverse impacts. Increased content of nitrogen (nitrates) in soil frequently leads to undesirable changes in vegetation composition and many plant species are endangered as a result of eutrophication in terrestrial ecosystem also. Chemical forms of nitrogen are most often of concern with regard to eutrophication because plants have high nitrogen requirements so that additions of nitrogen compounds stimulate plant growth. Ecosystems receiving more nitrogen than the plants require are called nitrogen saturated.

Ecological Effects

Many ecological effects can arise from stimulating primary production, but there are three particularly troubling ecological impacts:

- decreased biodiversity
- changes in species composition and
- toxicity effects

Decreased Biodiversity

When an ecosystem experiences an increase in nutrients, primary producers reap the benefits first. In aquatic ecosystems, species such as algae experience a population increase called an algal bloom. Algol bloom limits the sun light available to bottom-dwelling organisms and cause wide swings in the amount of dissolved oxygen in the water. When dissolved oxygen levels decline to hypoxic levels, fish and other marine animals suffocate, and ultimately die.

New Species Invasion

Eutrophication may cause competitive release by making abundant a normally limiting nutrient. This process causes shifts in the species composition of ecosystems. For instance, an increase in nitrogen might allow new, competitive species to invade and out compete original inhabitant species.

Toxicity

Some algal blooms, otherwise called "nuisance algae" or "harmful algal blooms" are toxic to plants and animals. Toxic compounds they produce can make their way up the food chain, resulting in animal mortality. Freshwater algal blooms can pose a threat to livestock. When the algae die or are eaten, neuro-and hepatotoxins are released which can kill animals and may pose a threat to humans.

Nitrogen can also cause toxic effects directly. When this nutrient is leached into groundwater, drinking water can be affected because concentrations of nitrogen are not filtered out. Nitrate (NO_3) has been shown to be toxic to human babies. This is because bacteria can live in their digestive tract that convert nitrate to nitrite (NO_2). NO_2 reacts with haemoglobin to form methaemoglobin, a form that does not carry oxygen. The body thus suffocates as its body receives insufficient oxygen.

Prevention and Reversal of Eutrophication

Eutrophication poses a problem not only to ecosystems, but to humans as well. Reducing eutrophication should be a key concern when considering future policy, and a sustainable solution for everyone, including farmers and ranchers, seems feasible.

While eutrophication, does pose problems, humans should be aware that natural run off, which cause algal blooms in the wild, is common in ecosystems and should thus not reverse nutrient concentrations beyong normal levels.

Cleanup measures have been mostly, but not completely, successful. Finish phosphorous removal measures started in the mid 1970s and have targeted rivers and lakes polluted by industrial and municipal discharges. These efforts have had a 90% removal efficiency.

PROBLEMS

1. Which are the gases majorly responsible for causing green house effect and how are they released into the atmosphere? What are the possible consequences of uncontrolled global warming?
2. How is ozone formed and depleted in nature? What are the consequences of the depletion of the O_3 layer in the atmosphere?
3. What are the pollutants responsible for causing acid rain? What are its deleterious effects on material and terrestrial ecosystems?
4. What are the different methods adopted to limit the levels of various air pollutants in the atmosphere?
5. Name important air pollutants and mention their effects on human life.
6. What is meant by the terms pollutant, contaminate, receptor, sink, reservoir?
7. Discuss the sources, harmful effects and methods of control of the following pollutants:
   (*a*) CO, (*b*) Particulate matter,
   (*c*) Photochemical oxidants.
8. Explain the harmful effects of air pollution with respect to:
   (*a*) Depletion of ozone layer (*b*) Photochemical smog
   (*c*) Green house effect.
9. Define pollution. List various sources of air pollution and discuss their effects on the environment.
10. Explain briefly
    (*a*) Acid rain, (*b*) Photochemical smog,
    (*c*) Greenhouse effect, (*d*) Ozone hole,
    (*e*) Radiation balance of earth.
11. Describe the sources and constituents of air pollution.
12. What is meant by green house effect? Describe two monitoring techniques used for green house effect.

13. What is the green house effect? How do pollutants affect this?
14. What is the effect of ozone depletion? How does it occur?
15. How does automobile exhaust contribute to air pollution?
16. How do you account for 'Ozone depletion'? Indicate its consequences.
17. Explain the sources and harmful effects of the following air pollutants.
(*a*) CO (*b*) SO_2
(*c*) Smoke (*d*) Particulate matter
18. Write detailed informative notes on the following:
(*a*) Earth's radiation balance: (*b*) Lapserate and temperature inversion.
19. How are air pollutants classified? What are the major effects of CO, SO_2, NO, Hydrocarbon and particulates on human health.
20. What are the sources of NO and SO_2 global atmosphere? Give an account of the global efforts to control the release of these two pollutants into the atmosphere by anthropogenic source.
21. "Carbon dioxide, a non-pollutant, is perhaps the single most important environmental question facing us at present." Discuss in terms of the green house effect.
22. Discuss briefly ozone hole and its effect on human health.
23. What are the primary pollutants? Discuss their sources and relative contribution to air pollution.
24. How can internal combustion engine be modified to make auto-exhausts free from pollutants?
25. How do you account for Pb in particulate matter? Can you suggest a corrective step for making air free from Pb?
26. Write in brief about water pollution.
27. What are the sources and cause of water pollution?
28. Describe in details about types of water pollutants.
29. Explain in details about Sewage Treatment.
30. Write short notes on Green Chemistry and Eutrophication.

CHAPTER

4

METAL FINISHING

4.1 INTRODUCTION

The term metal finishing is of long standing acceptance and remains in common use despite some inaccuracy. Metal is no longer the dominating material of engineering construction and metal is no longer the main coating material used in finishing operations. Finishing is appropriate because the last operation applied to an engineering component, with only a few exceptions often involving heat treatment, is that the surface relating to its appearance.

A generic name has become common, as a consequence, and surface engineering has come to imply the design and manufacture of a surface for engineering property advantage. Therefore, the term metal finishing may be defined as:

"The deposition of coating (metallic or non-metallic) onto a metallic or non-metallic substrate, or the treatment of metallic substrate by the use of a physical, chemical or mechanical process yielding a thin film, to enhance the appearance, function or performance of a product and to give added value and promote marketability".

Finishing necessarily includes a large range of operations—chemical, physical and mechanical and other specialist bodies respond to industrial sector needs, *e.g.*, galvanisers, metal sprayers, painters, vitreous enamellers etc. The Institute of Metal Finishing (IMF) embraces all these technologies but tends to emphasise particular branches.

4.2 DESIGNING A COATING SYSTEM

Following five stages are applied for the design of a coating system:

(*i*) Surface preparation for cleaniness
(*ii*) Surface pretreatment
(*iii*) Application of a first 'strike' or 'primer' layer on the substrate
(*iv*) Application of the main underlayer to influence long term service
(*v*) Application of the final top coat giving the full surface properties.

4.2.1 Surface Preparation

Considering the term 'A coating is only as good as the surface on which it is laid' the substrate surface should always be clean, free of corrosion product, free of old coating, blemish free, smooth level and preferably bright. Surface preparation includes the following steps:

(*i*) Cleaning of metals by organic solvent, degreasing or by aqueous alkali detergent treatment. These processes aided higher temperatures and scrubbing action including ultrasonic agitation to remove physically attached soils.

(*ii*) Pickling of metals, usually, but not exclusively, by acids, to remove chemically attached soil such as tarnish, hot scale and corrosion products.

(*iii*) Sometimes, when these two methods are not useful, typically re-coating large engineering assemblies other physical and mechanical methods can be used.

(*iv*) Flame cleaning of metals where an oxy-acetylene flame utilises heat and jetting to expand and 'blow-off' soil.

(*v*) Shot blasting, with air and water medium, to remove gross scale. This process is widely used for bridges, buildings etc.

(*vi*) Wire power tool abrasion is an easy DIY technique allowing continual inspection of the surface.

4.2.2 Surface Pretreatment

When the surface is clean a second step called surface pretreatment must be taken as soon as possible because surface deterioration especially by polluted atmospheres, can be rapid. Pretreatment processes includes the following:

(*i*) Corrosion control requires chemical treatments to stabilise the surface against deterioration before the first real coating layer is applied.

(*ii*) Conversion coatings or etch-primer types of treatment are used. The commonest conversion coatings are phosphating and chromating but others such as oxalating have found specialist uses.

(*iii*) Chemical etching can be employed to develop various levels of matt finish which are then perpetuated through thin overlay coatings.

(*iv*) Polishing was traditionally carried out manually as a labour intensive operation. Now-a-days chemical and electrochemical polishing is preferred.

(*v*) Mechanical modification, usually to develop 'brushed' finishes can be achieved using steel wire brushes and roll embossing where the roll have been brushed or shot blasted.

4.2.3 Strike

Initially, a special plating deposite called a "strike" may be used to form a very thin (less then 0.5 mm thick) plating with high quality and good adherence to the substrate. This serves as a foundation for subsequent plating processes. A strike use a high current density and a bath with a low ion concentration. The striking method is also used in combination with the plating of different metals. If it is desirable to plate one type of deposite onto a metal to improve corrosion resistance but this metal has inherently poor adhesion to the substrate, a strike can be first deposited that is compatible with both.

One example of this situation is the poor adhesion of electrolytic nickel on zinc alloys, in which case a copper strike is used, which has good adherence to both.

4.3 ELECTROPLATING

The terms electroplating or electrodeposition or plating commenced in 1839-1840 in the UK and Russia using gold and silver on copper or brass for tableware, ornaments etc. Electroplating involves the coating of an electrically conductive object with a layer of metal using electrical current. Usually, the process is used to deposite an adherent surface layer of a metal having some desired property, *e.g.*, abrasion and wear resistance, corrosion protection, lubricity etc. onto a substrate lacking that property. It is also used to build up thickness on undersized parts.

The process used in electroplating is called Electrodeposition and is analogous to a galvanic or electrochemcial cell acting in reverse. The part to be plated is the cathode of the circuit while metal to be plated on the part is act as anode. Both of these components are immersed into a solution containing one or more metal salts as well as other ions that permit the flow of electricity. A rectifier supplies a direct current to the cathode causing the metal ions in solution to lose their charge and plate out on the cathode. As the electrical current flow through the circuit, the anode slowly dissolves and replenishes the ions in the bath.

On the contrary, other electroplating processes use a non-consumable anode such as lead, where ions of the metal to be plated must be periodically replenished in the bath as the plate forms out of the solution.

4.3.1 Procedure of Electroplating

The anode and cathode in the electroplating cell are connected to external supply of direct current, a battery, or more commonly a rectifier. The anode is connected to the positive terminal of the supply, and the cathode (metals to be plated) is connected to the negative terminal as represented in Fig. 4.1.

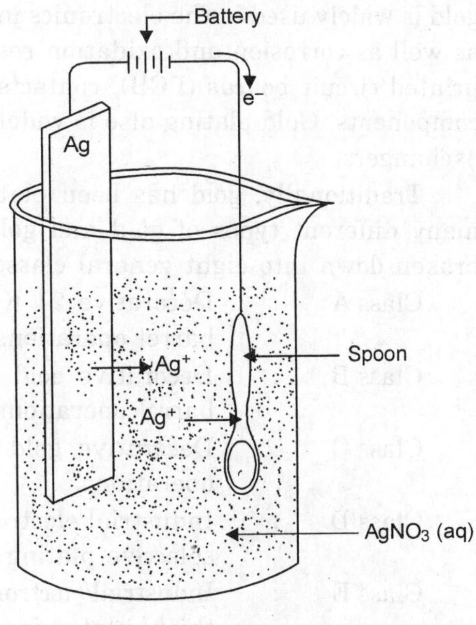

Fig. 4.1: Diagramatic illustration of silver electroplating.

When the external power supply switched on, the metal at the anode is oxidized from the zero valency state to form cations with a positive charge.

These cations associated with the anions in the solution. The cations are reduced at the cathode to deposite in the metallic, zero valency state. For examples: Cu is oxidized from a copper anode to Cu^{2+} by losing two electrons. The Cu^{2+} associates with the anion SO_4^{2-} in the solution to form $CuSO_4$. At the cathode, the Cu^{2+} is reduced to metallic Cu by gaining two electrons, resulting in the transfer of Cu from the anode source to a plate covering the cathode.

The plating is mostly a single metallic element, not an alloy. However, some alloys can be electrodeposited, notably brass and solder.

Many plating baths include cyanides of other metals *e.g.*, potassium cyanide, in addition to cyanides of the metal to be deposited. These free cyanides facilitate anode corrosion, help to maintain a constant metal ion level and contribute to conductivity. Additionally, non-metal chemicals such as carbonates and phosphates may be added to increase conductivity.

If plating is not desired on certain areas, stop-offs are applied to prevent the bath from coming in contact with the substrate. The typical stop-offs are tape, foil, waxes and lacquers.

4.3.2 Brush Electroplating

In brush electroplating process, localized areas or entire items are plated using a brush saturated with plating solution. The brush is made up of stainless steel body wrapped with a cloth material that both holds the plating solution and prevents direct contact with the item being plated, is connected to the positive side of a low voltage direct current power source, and the item to be plated connected to the negative. The operator dips the brush in plating solution then applies it to the item, moving the brush continually to get an even distribution of the plating material.

The brush acts as the anode, but does not contribute any plating material. Although sometimes the brush is made from or contains the plating material in order to extend the life of the plating solution.

4.3.3 Electroplating of Gold

Gold plating was used primarily for decorative purposes in jewellery and flatware. Currently, gold is widely used in the electronics industry because of its good electrical contact properties as well as corrosion and oxidation resistance. Typical applications for gold plating include printed circuit boards (PCB), contacts, connectors, transistor bases, and integrated circuit components. Gold plating also is widely used in the chemical industry for reactors and heat exchangers.

Traditionally, gold has been plated from potassium gold cyanide solutions, although many different types of gold and gold alloys are available. However, gold plates can be broken down into eight general classes:

Class A	:	Decorative 24 K gold flash (2-4 millionths thick) plated in rack and barrel operations.
Class B	:	Decorative gold alloy flash (2-4 millionths thick) plated in rack and barrel operations.
Class C	:	Decorative gold alloy flash (20-40 millionths thick) plated in rack operations.
Class D	:	Industrial/electronic high purity soft gold plated in rack, barrel and selective plating operations.
Class E	:	Industrial/electronic hard, bright and heavy 99.5% gold (20-200 millionths thick) plated in rack, barrel, and selective plating operations.
Class F	:	Industrial/electronic gold alloy heavy (20-400 millionths thick) plated in rack and selective plating operations.
Class G	:	Refinishing, repair and general; deposites are pure bright alloys (5-40 millionths thick) plated in rack and selecting plating operations.
Class H	:	Miscellaneous including electro forming of gold and gold alloys, statuary, and architectural applications.

Common Bath Solutions

The four general groups of gold plating solutions are: alkaline gold cyanide, neutral gold cyanide, acid gold cyanide and non-cyanide solutions (generally sulfite based).

- Alkaline gold cyanide : Class A, B, C, D occasionally F, G, and H.
- Neutral gold cyanide : This is usually used to achieve high purity gold plated for class D and G.
- Acid gold cyanide : This is used for bright, hard gold and gold alloy plating. It is used occasionally for class B, C, E, F and G.
- Non-cyanide gold : Occasionally, it is used for class A, B, C, D, F, G and H.

Besides above following alternative gold plating solutions are being used:

- Gold sulfite
- Cobalt-Hardened gold (No free cyanide)

The gold plating cell used in industries is as illustrated in Fig. 9.4.2.

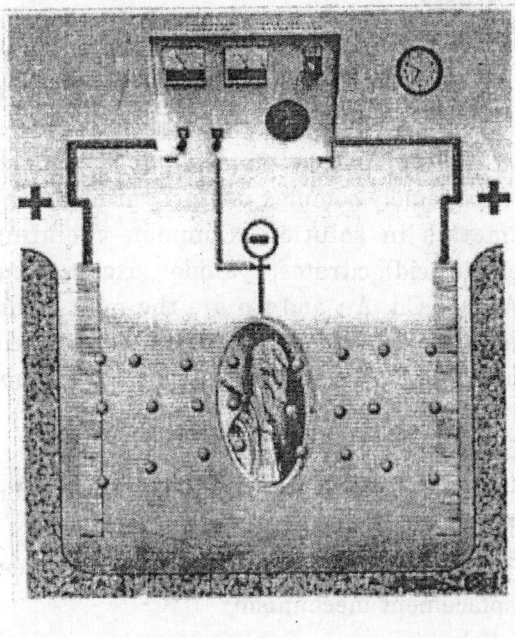

Fig. 4.2: Diagramatic illustration of gold plating cell

4.3.4 Effect of Plating Variables on Nature of Electroplating

The current density, *i.e.*, an amperage of the electroplating current divided by the surface area of the part, in electroplating process strongly influences the deposition rate, plating adherence, and plating quality. The current density can vary over the surface of a part, as outside surfaces will tend to have a higher current density than inside surfaces, *e.g.*, holes, bores etc.

The higher the current density, the faster the deposition rate will be, although there is a practical limit enforced by poor adhesion and plating quality when the deposition rate is too high.

4.3.5 Applications of Electroplating

Following are the main applications of electroplating:

- Decorative/aesthetic – Cosmetic marketable appearance
- Functional/engineering – Corrosion and water resistance
- Electronic circuitry – Printed circuit board (PCB) connectors and contacts
- Repair/reclamation – Refurbishment of worn parts, recovery of tolerance
- Manufacturing – Electroforming
- Economic – Cheaper (*e.g.*, steel) substrate expensive coating (*e.g.*, tin)

4.4 ELECTROLESS PLATING

Electroless plating is a growing segment of metal finishing, especially in the electronics industry. The name electroless plating itself means electroless current deposition. In electroless plating, metals are deposited onto the surface of a part without the use of electricity as a source of electrons. Instead, the bath solution supplies the electrons for the deposition reaction. These baths are extremely complex using a variety of chelating and/or complexing agents that hold the metals in solution. Common chelating agents include EDTA (ethylenediamine tetra acetic acid), citrates, cyanides, oxalates and 1, 2 diaminocyclohexane tetra acetic acids (DCTA). Ni, Cu, Au and Co are the most common metals plated in this process. Deposition rates are controlled by the amount of reducing agent present and the type of chelating agent used. Common electroless metals are: copper, nickel, silver and gold.

4.4.1 Procedure of Electroless Plating

Electroless plating doesn't require a continuous conductive path for plating. Onto-plating can be done straight onto plastic and other substrates.

Electroless plating is generally carried out by two processes, *viz.*,

(*i*) Galvanenic displacement mechanism

(*ii*) Autocatalytic mechanism

Galvanic Displacement Mechanism

This technology has been known for many years and was exploited in the development of batteries. The battery generally has electrodes that are made of different metals, *i.e.*, different electrode potentials and an electrolyte *i.e.*, a sea of ions. The technology has also been exploited for metal plating such as a process for silvering mirrors.

Galvanic displacement relies on relative positions in electrochemical series. Standard electrode potentials of some elements are as given on next page:

Iron	–0.44
Nickel	–0.24
Tin	–0.14
Lead	–0.13
Hydrogen	+0.00
Copper	+0.34
Iodine	+0.54
Silver	+0.80
Gold	+0.80

The order represented above shows the tendency of one metal to reduce the ions of any other metal below it in the series, *e.g.*, silvering of mirrors.

Autocatalytic Mechanism

Autocatalytic electroless plating became common around 1965, when plating on plastic objects achieved a significant development. Autocatalytic process requires the presence of a catalyst. Palladium and some other noble metals are being used as catalyst.

Autocatalytic requires the employment of a catalytic metal to kickstart and aid sustainment of the reaction.

4.4.2 Electroless Plating of Copper on PCB and Nickel

In electroless copper plating, the Rochelle-salt bath is commonly used to plate plastic objects cosmetically, and the EDTA bath to plate PCB. Complexing agents are essential to produce good plating. Typical copper loading is 2 to 3 gm per litre of solution. Typical copper plating solution, *e.g.*, Shipley 3350 will contain chemicals such as $CuSO_4$, HCl, EDTA and NaOH.

Following four phases/stages are being applied during electroless plating of copper plating on PCB:

(*i*) Conditioning
(*ii*) Etching
(*iii*) Activating
(*iv*) Deposition

4.4.2.1 Conditioning

This ensures that all surfaces are clean and free from grease or metal contaminants usually done by immersing the substrate in a chemical agent. Deburring of holes also done with rotating brushes.

4.4.2.2 Etching

It is also known as pickling to differentiate it from etching of the copper tracks. Prepares the surface for deposition and removes any deformities left from previous stages. Heated sulphuric acid and H_2O_2 solutions are commonly used here.

4.4.2.3 Activating

Activating is used to create a chemically reducing surface. Traditionally, there are two processes for activating being used:

(*i*) first of all the laminate is sensitized with stannous (tin) ions on the board.

(*ii*) secondly, the board is seeded by dipping in acidic palladium chloride. Palladium ions reduced to a colloidal state during reaction with stannous ions.

4.4.2.4 Deposition

In the process the board is dipped into a solution containing copper ions. The chemically reducing surface of the board (colloidal palladium) causes a reaction and copper is deposited onto the board. Typical thickness is 1 to 5 microns. 25 microns is possible through for high build applications.

A similar process can be performed for the deposition of copper onto Nickel layer.

During the electroless plating stage, for the inter connection of the multiple layers by the means of vias, a resist is also used on the top and bottom layer to prevent plating over unwanted areas.

Copper PCB is used as a cathode (negative charge). Positively charged copper ions from copper sulfate are attracted to it, subsequently SO_4^{-2} ions are attracted to anode. The diagrammatic representation of above is as illustrated in Fig. 4.3.

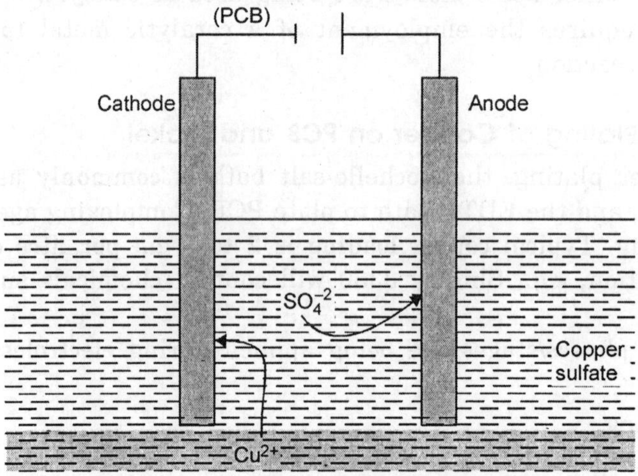

Fig. 4.3: Diagramatic illustration of electroplating of Cu on PCB

Use of resists in copper plating through hole interconnects is as given below in the figure 4.4.

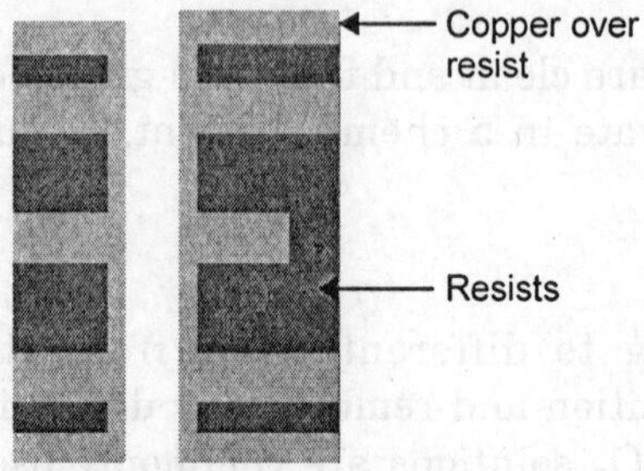

Fig. 4.4: Diagramatic illustration of use of resists in copper plating

A completed metallised multilayer board is as illustrated in Fig. 4.5.

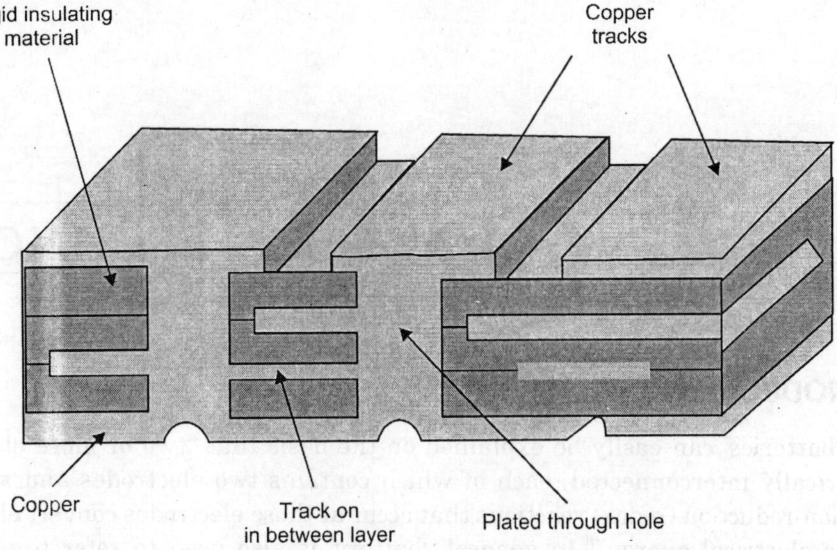

Fig. 4.5: Diagramatic illustration of completed metallised multilayer board

4.4.3 Effect of Electroless Plating Variables on Electroless Plating

Electroless plating is temperature dependent. Deposition is accelerated by the use of high temperature. The deposit may contain reduction agent residues such as phosphorous from hypophosphite; this can be exploited by post deposition heat treatment to substantially improve hardness and wear resistance through precipitation of phosphides in nickel.

4.4.4 Applications of Electroless Plating

Being advantageous in many respects such as uniform precipitation coating, hardness, abrasion resistance, chemical resistance and corrosion resistance, electroless plating became increasingly common especially in precision machinery industries.

PROBLEMS

1. Explain the term Metal finishing.
2. Illustrate the term electroplating and electroless plating.
3. Give in details about the application of electroplating and electrolessplating.
4. Explain the following:

(*i*) Surface preparation	(*ii*) Strike
(*iii*) Surface pretreatment	(*iv*) Electroplating of gold
(*v*) Silver electroplating	(*vi*) Electroplating of Cu on PCB and Ni

CHAPTER

5

BATTERY TECHNOLOGY

5.1 INTRODUCTION

The term 'batteries' can easily be explained on the basis that "two or more electrochemical cells, electrically interconnected, each of which contains two electrodes and an electrolyte. The oxidation-reduction (redox) reactions that occur at these electrodes convert electrochemical energy into electrical energy." In general, 'battery' is also used to refer to a single cell.

Fig. 5.1 shows the most simple design of an electrochemical cell.

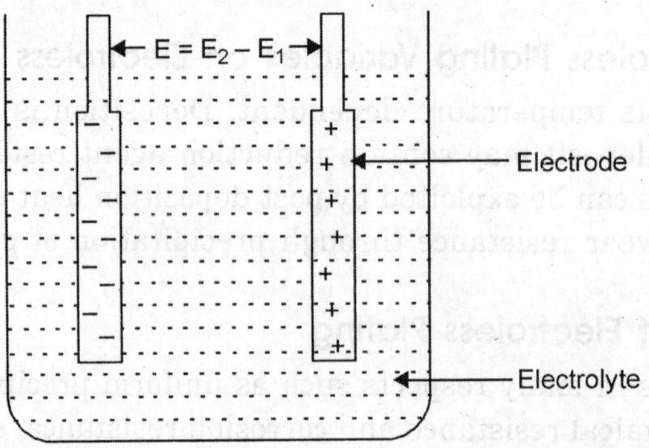

Fig. 5.1: Diagrammatic illustration of an electrochemical cell

Two electrodes are immersed in an electrolyte resulting in a voltage difference E at their terminals, which is composed as demonstrated by Fig. 5.2, of voltage differences, namely the voltage differences from the electrodes *Vs* the electrolyte.

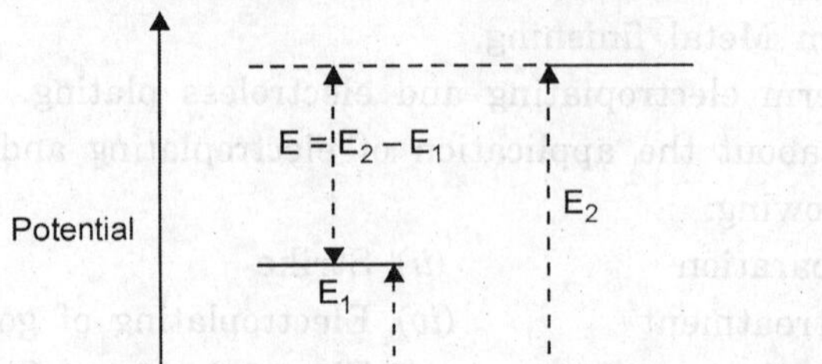

Fig. 5.2: Voltage differences from the electrodes *Vs* the electrolyte

5.2 CHARACTERISTICS OF BATTERIES

The batteries have been developed for a wide range of applications using a variety of different technologies, resulting in a wide range of available performance characteristics. It is essential to describes the main parameters which are used to characterise cell performance.

5.2.1 The Charge/Discharge Curve

The measured terminal voltage of any battery will vary as it is charged and discharged. The charge and discharge curve of various batteries are diagrammatically represented in Fig. 5.3.

The Mid-point Voltage (MPV) is the nominal voltage of the cell, and is the voltage that is measured when the battery has discharged 50% of its total energy where as the measured cell voltage at the end of its operating life is called the End of Discharge Voltage (EODV). Sometimes refered as End of Life Voltage (EOLV). The MPV and EODV is illustrated in Fig. 5.4.

When the battery is peak charged, the actual cell voltage will be higher than the MPV, and when nearing the EODV point, the cell voltage will be less than the MPV.

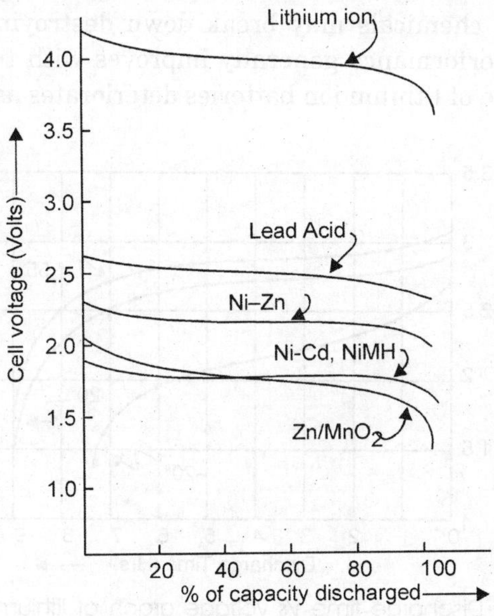

Fig. 5.3: Charge/discharge curve of various batteries

The Fig. 5.3 clearly shows that lithium ion have a fairly flat discharge curve while others such as lead acid have a pronounced slope. The power delivered by cells with a sloping discharge curve falls progressively throughout the discharge cycle. This could give rise the problems for high power applications towards the end of the cycle. For low power applications which need a stable supply voltage, it may be necessary to incorporate a voltage regulator if the slope is too steep.

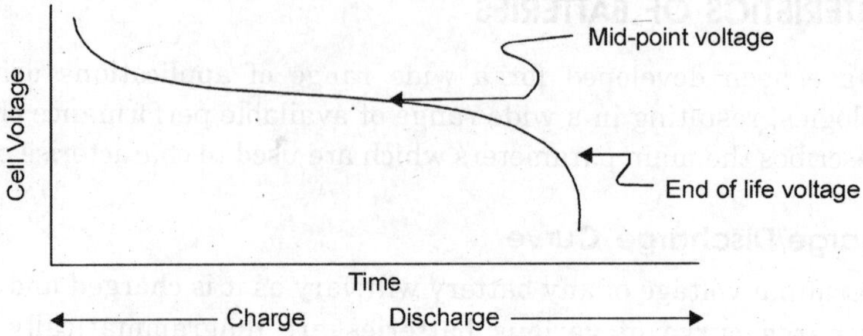

Fig. 5.4: Battery charge and discharge curve

In Fig. 5.3, the X-axis shows the cell characteristics normalised as a percentage of cell capacity so that the shape of the graph can be shown independent of the actual cell capacity. If the X-axis was based on discharged time, the length of each discharge curve would be proportional to the nominal capacity of the cell.

5.2.2 Temperature Characteristics

The performance of battery can change dramatically with temperature. At the lower extreme the electrolyte may freeze setting a lower limit on the operating temperature, while at the upper extreme the active chemicals may break down destroying the battery. In between these limits the battery performance generally improves with temperature. The figure 5.5 shows how the performance of lithium ion batteries deteriorates as the operating temperature decreases.

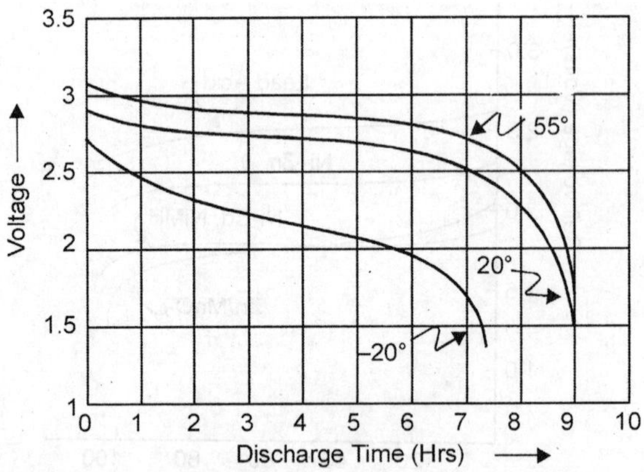

Fig. 5.5: Discharge time *Vs* voltage graph of lithium batteries

5.2.3 Self Discharge Characteristics

The self discharge rate is measured of how quickly a cell will lose its energy while setting on the shelf due to unwanted chemical actions within the cell. The rate depends on the cell chemistry and the temperature.

The following shows the typical shelf life for some primary cells:

Leclanche (Zinc Carbon) – 2 to 3 years

Alkaline – 5 years

Lithium – 10 years or more

Typical self discharge rates for common rechargeable cells are as follows:

Lead Acid – 4% to 6% per month

Nickel Cadmium – 15% to 20% per month

Nickel Metal Hydride – 30% per month

Lithium – 2% to 3% per month.

The rate of unwanted chemical reactions which cause internal current leakage between the +ve and –ve electrodes of the cell, like all chemical reactions, increases with temperature thus increasing the battery self discharge rate. The Fig. 5.6 shows typical self discharge rates for a lithium ion battery.

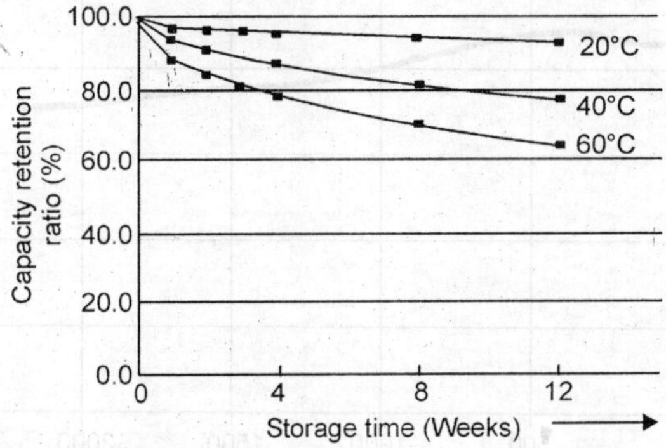

Fig. 5.6: Diagrammatic representation of a typical self discharge graph for lithium ion battery

5.2.4 Energy Density (By weight and volume)

The energy density of a battery is generally expressed in two ways; as illustrated in Table 5.1. The gravimetric energy density of a battery is a measure of how much energy a battery contains in comparison to its weight, and is typically expressed in watt-hours/kilogram (W-hrs/kg).

The volumetric energy density of a battery is a measure of how much energy a battery contains in comparison to its volume, and is typically expressed in (W-hrs/liter).

Table 5.1: Energy density comparison

Cell Type	*Ni-MH*	*Ni-Cd*	*Li-Ion*
Gravimetric Density (W-hrs/kg)	55	50	90
Volumetric Density (W-hrs/lt)	180	140	210

The above table clearly shows that the gravimetric density of Li-Ion is almost doubling the density of Ni-Cd and Ni-MH. If the battery weight is kept the same, the run time will double if Li-Ion batteries are used.

5.2.5 Cycle Life

This is one of the key cell performance parameters and gives an indication of the expected working lifetime of the cell. The cycle life is defined as the number of cycles a cell can perform before its capacity drops to 80% of its initial specified capacity. The graphic representation of cycle life of a battery at room temperature is given in Fig. 5.7. Which clearly indicates that the cell does not die suddenly but continues a slow deterioration which will be almost imperceptible to the user. At the end of the specified cycle life the cell will continue to function normally except that its capacity will be significantly less than it was when it was new.

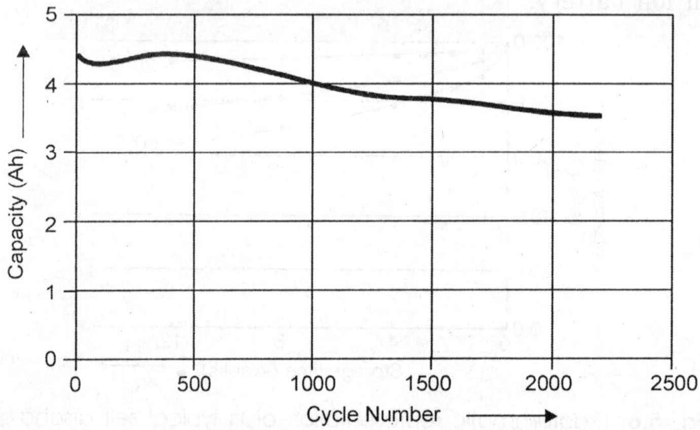

Fig. 5.7: Cycle life at room temperature

5.2.6 Peukert Equation

The Peukert equation is a convenient way of characterising cell behaviour and of quantifying the capacity offset in mathematical terms.

This is an emperical formula which approximates how the available capacity of a battery changes according to the rate of discharge,

$$C = I^n T$$

Where, 'C' is the theoretical capacity of the battery expressed in amp hrs.

'I' is the current

'T' is the time and

'n' is the Peukert number, a constant for a given battery.

The above equation shows that at higher currents there is less available energy in the battery.

The Peukert number is directly related to the internal resistance of the battery. Higher current means more losses and less available capacity. The value of Peukert number indicates how well a battery performs under continuous heavy current. A value close to 1 indicates that the battery performs well. The higher the number, the more capacity is lost when the battery is discharged at high current. Peukert curve is illustrated in Fig. 5.8.

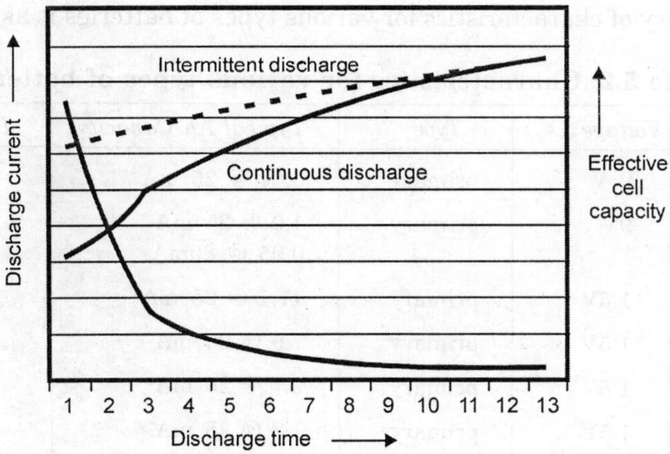

Fig. 5.8: Peukert curve

The peukert curve shows that the effective battery capacity is reduced at very high continuous discharge rates. For lead acid battery the Peukert number is typically between 1.3 and 1.4.

5.2.7 Ragone Plot

The Ragone plot is another way of characterising the reduction in effective capacity or capacity offset as the discharge rate or power delivered is increased.

The Fig. 5.9 is also useful for demonstrating the superior gravimetric energy density of lithium ion cells.

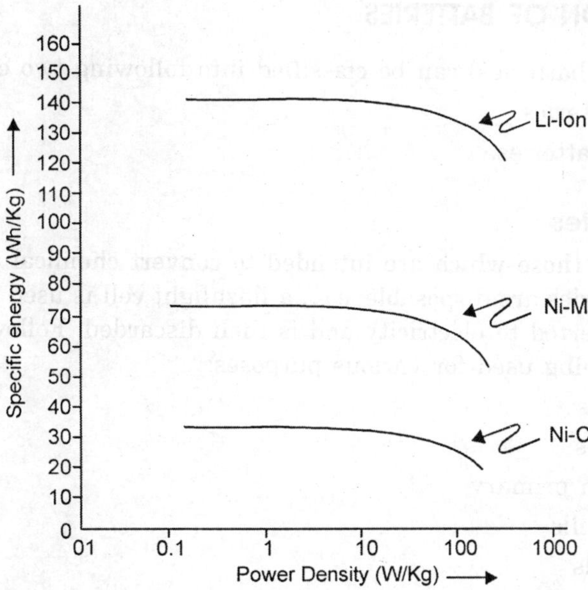

Fig. 5.9: Ragone plot

Similar graph can be plotted for the volumetric energy density.

A brief summary of characteristics for various types of batteries is as given in table 5.2.

Table 5.2: Characteristics for various types of batteries

Type	*Voltage (OC)*	*Type*	*Typical Ah Capacity*	*R Internal (Ω)*
9V Alkaline	9 V	primary	0.60 @ 25 mA	2
9V Lithium	9V	primary	1.0 @ 25 mA 0.95 @ 80mA	18
Alkaline D	1.5V	primary	17.1 @ 25 mA	0.1
Alkaline C	1.5V	primary	7.9 @ 25 mA	0.2
Alkaline AA	1.5V	primary	2.7 @ 25 mA	0.4
Alkaline AAA	1.5V	primary	1.2 @ 25 mA	0.6
Ni-Cd D	1.3 V	secondary	4.0 @ 800 mA 3.5 @ 4 mA	0.009
Ni-MH AAA	1.2 V	secondary	0.55 @ 200 mA	–
Ni-MH AA	1.2 V	secondary	1.3 @ 200 mA	–
Ni-MH C	1.2 V	secondary	3.5 @ 200 mA	–
Ni-MH D	1.2 V	secondary	7.0 @ 200 mA	–
Ni-MH 9V	8.4 V	secondary	0.13 @ 200 mA	–
Lithium ion	3.7 V	secondary	0.76 @ 200 mA	–

5.3 CLASSIFICATION OF BATTERIES

Electrochemical cells (batteries) can be classified into following two categories:

(*i*) Primary batteries

(*ii*) Secondary batteries.

5.3.1 Primary Batteries

Primary batteries are those which are intended to convert chemical energy into electrical energy. Primary batteries are disposable, *e.g.*, a flashlight cell is used until all the chemical energy has been converted to electricity and is then discarded. Following are the primary batteries which are being used for various purposes:

(*i*) Zinc/Air cells

(*ii*) Lithium cells

(*iii*) Lithium iron primary

(*iv*) Leclanche cells

(*v*) Alkaline cells

(*vi*) Mercury oxide cells

(*vii*) Aluminium/Air cells

(*viii*) Magnesium–copper chloride reserve.

Here under this heading, we will consider only first two, *i.e.*, Zinc/Air cells and Lithium cells as an example.

5.3.1.1 *Zinc-air Cells*

The Zinc air cell fits into the alkaline cell category because of its electrolyte. It also acts as a partial fuel cell because it uses the O_2 from air as the cathode. Actually, oxygen is let in to the cathode through a hole in the battery and is reduced on a carbon surface. In Zinc-Air cell, the amalgamated zinc powder and electrolyte is act as anode whereas Oxygen (O_2) is act as cathode. Potassium hydroxide (KOH) is also used as electrolyte.

The metal oxide reduces, the zinc which becomes oxidized. As a result electric current is produced. A familiar example is the old mercury Oxide/Zinc battery used for hearing aids.

The electrochemical reactions involved in the above said Zinc-air battery is as given below:

Location	*Half cell reactions*	*Voltage*
Anode	$Zn^{2+} + 2OH^- \rightarrow Zn(OH)_2$	1.25
Cathode	$1/2O_2 + H_2O + 2e \rightarrow 2OH^-$	0.4
Overall	$2Zn + O_2 + 2H_2O \rightarrow 2Zn(OH)_2$	1.65

As we know that KOH is used as electrolyte, which is hygroscopic in nature. Because of this life of the cell will be reduced. KOH used as electrolyte react with CO_2 in the air to form alkali carbonates. For this reason large Zinc-air batteries usually contain a higher volume of CO_2 absorbing material, *i.e.*, calcium oxide flake, than battery components. This can cancel out the huge increase in energy density gained by using the air electrode. This cell has the additional benefits of being environmentally friendly at a relatively low cost. The Zinc-air battery is applicable in hearing aids, pagers, electric vehicles, etc.

5.3.1.2 *Lithium Cells*

Lithium battery chemistry comprise a number of cell designs that use lithium as the anode. Lithium is gaining a lot of popularity as an anode, due to the highest standard potential of all the metals, at over 3 volts, means, lithium has the highest energy density. Lithium cells exists because of its stability and low reactivity with a number of electrodes and non-aquous electrolytes. The most common electrolytes are organic liquids with the notable exceptions of thionylchloride (SO_2Cl_2) and sulfuryl chloride (SO_2Cl_2).

Lithium cells can be grouped into three general categories, *viz.*,

(*i*) Liquid Cathode

(*ii*) Solid Cathode and

(*iii*) Solid Electrolyte.

Solid cathode lithium cells

These cells cannot be used in high-drain applications and don't perform as well as the liquid cathode cells in low temperatures. However, they don't have the same voltage delay and the cells don't require pressurization. Solid cathode lithium cells are generally used for memory backup, watches, portable electronic devices, etc.

Lithium -MnO$_2$ battery

Lithium-MnO$_2$ battery account for about 80% of all primary lithium batteries. This is only because of their low cost. The cathode used is a heat-treated MnO$_2$ and the electrolyte a mixture of propylene carbonate and 1,2–dimethoxyethane. The half reactions occured are:

Anode : $Li \rightarrow Li^+ + e^-$

Cathode : $Mn^{IV}O_2 + Li^+ + e^- \rightarrow Mn^{III}O_2(Li^+)$

Overall : $Li + Mn^{IV}O_2 \rightarrow Mn^{III}O_2(Li^+)$

At lower temperatures and in high-rate discharge, the LiSO$_2$ cell performs much better than the LiMnO$_2$ cell. At low rate discharge and higher temperatures, the two cells perform equally well, but LiMnO$_2$ cell has the advantage because it does not require pressurization.

5.3.2 Secondary Batteries

Secondary batteries are those which are intended to store electrical energy as chemical energy and then resupply electricity on demand. Secondary batteries are also called as rechargeable batteries. Following are the secondary batteries, being used for various purposes:

(*i*) Nickel-Metal hydride batteries
(*ii*) Nickel-Cadmium batteries
(*iii*) Lead-Acid batteries
(*iv*) Sodium-Sulphur batteries
(*v*) Nickel-Sulphur batteries
(*vi*) Lithium ion cells
(*vii*) Manganese-Titanium (lithium) cells
(*viii*) Rechargeable alkaline manganese cells
(*ix*) Niekel Zinc cells
(*x*) Iron Nickel cells
(*xi*) Iron-Air cells
(*xii*) Iron-Silver cells
(*xiii*) Redox (liquid Electrode) cells

Here under this heading, we will consider only Nickel- Metal hydride battery as an example.

5.3.2.1 Nickel-Metal Hydride Battery

Nickel/Metal hydride (NiMH) cell is a hybride of the NiCd and NiH$_2$ cells. Previously, this battery was not available for commercial use because, although hydrogen has wonderful anodii qualities, it requires cell pressurization.

In this battery, the anodes used are complex alloys containing many metals, such as an alloy of V, Ti, Zr, Ni, Cr, Co, Fe etc. where as Nickel Oxyhydroxide is being used as cathode. Potassium hydroxide (6M KOH) is used as an electrolyte. Following are the cell reactions:

Location	*Reactions*	*Voltage*
Anode	$MH + OH^- \rightarrow M + H_2O + e^-$	0.83
Cathode	$NiOOH + H_2O + e^- \rightarrow Ni(OH)_2 + OH^-$	0.52
Overall	$NiOOH + MH \rightarrow Ni(OH)_2 + M$	1.35

The underlying chemistry of these alloys and reasons for superior performance are not clearly understood, and the compositions are determined by emperical testing methods. The interesting feature of these alloys is that some metals absorb heat when absorbing hydrogen *and some give off heat when absorbing hydrogen*. Both of these are bad for a battery, since we would like the hydrogen to move easily in and out without any energy transfer. The successful alloys are all mixtures of exothermic and endothermic metals to achieve this.

Such type of batteries are used for the purpose of cellular phones, cam corders, emergency backup lighting, power tools, laptops, portable electric vehicles etc.

5.4 RESERVE BATTERY (LITHIUM-VANADIUM PENTOXIDE)

Li-V_2O_5 battery is somewhat different from other lithium systems as it is only applied for *military purposes as a so called 'reserve cell'* however in great quantities.

It is hermetically sealed in a stainless steel container with a glass duct while the electrolyte is situated in a fused glass container within the cell. It must be mechanically destroyed for operation, *i.e.*, by an outside impulse.

The cathode material consists of vanadium pentoxide (V_2O_5) mixed with 10% graphite. Lithium arsenic hexafluoride ($LiAsF_6$) and lithium-borate tetrafluoride ($LiBF_4$), dissolved in propylene carbonate and methyl formate or dimethoxy ethane is used as an electrolyte.

Due to the separate storage of anode, cathode and electrolyte, its life is nearly unlimited. Its main application is the power supply to ground and sea mines and to grenades.

Discharge is made over two voltage plateaus of first 3.4 V and then decreasing to 2.3 V, although the discharge mechanisms are still not very clear. The level of the second lower voltage plateau depends much more on the ambient temperature and on the discharge current, a point which must be taken into account during application.

The discharge gradients of a 185 mA cell of the system Li-V_2O_5 reserve cell is as illustrated in Fig. 5.10.

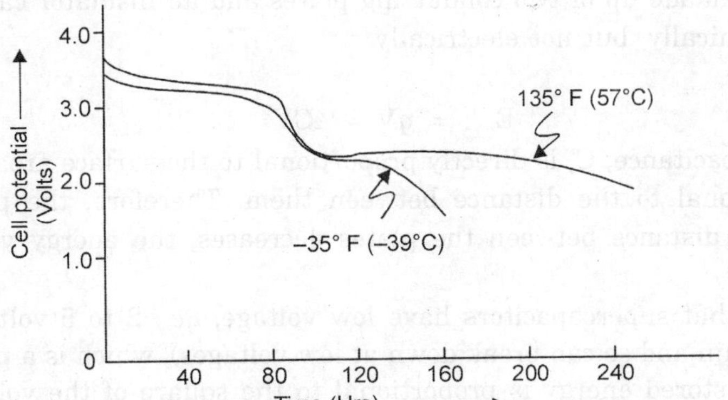

Fig. 5.10: Discharge gradients of a 185mA cell of the Li-V_2O_5 reserve cell at 3.25 k ohm (1 mA at 3.25 V) load

In comparison with thionylchloride ($SOCl_2$) and sulphur dioxide as illustrated in Fig. 5.11, some more, small plateaus can be recognised which, compared with the strong drop after about 50% of discharge, are of minor importance. The energy density at about 200 Wh/kg per weight and 600 Wh/l per volume is the average rate of the other lithium systems as well as the range of operation from –40°C to +60°C.

The above said system is constructed by HONEYWELL in sizes of 100 mAh to 30 Ah. For the construction of button cells originally produced with this system, other systems are being adopted now a days.

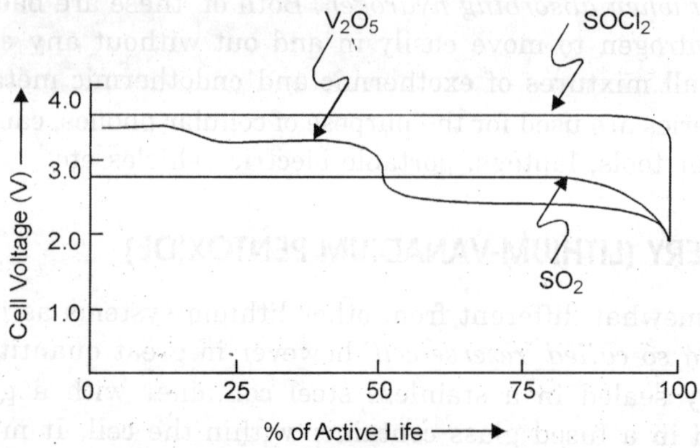

Fig. 5.11: Comparison of discharge gradients between Li-$SOCl_2$, Li-V_2O_5 and Li-SO_2

It also must be said that other lithium systems, lithium thionyl chloride and Lithium-sulphur dioxide, have gained importance also for reserve cells and are slowly replacing the vanadium pentoxide.

5.5 SUPERCAPACITORS

Unlike batteries, which store energy chemically, capacitors store energy as an electrostatic field. A battery is known for storing a lot of energy and little power, whereas a capacitor can provide large amounts of power, but low amounts of energy.

A capacitor is made up of two conducting plates and an insulator called the dielectric, which conducts ionically, but not electrically.

In a capacitor,

$$E_{cap} = qV = \frac{1}{2}CV^2$$

Where the capacitance, C, is directly proportional to the surface area of the plates and inversely proportional to the distance between them. Therefore, the plate surface area increases and the distance between the plates decreases, the energy you can store in a capacitor increases.

As we know that supercapacitors have low voltage, *i.e.*, 2 to 5 volts (remember that insulator is ultrathin and so can break down at low voltages), which is a problem for energy storage, since the stored energy is proportional to the square of the voltage.

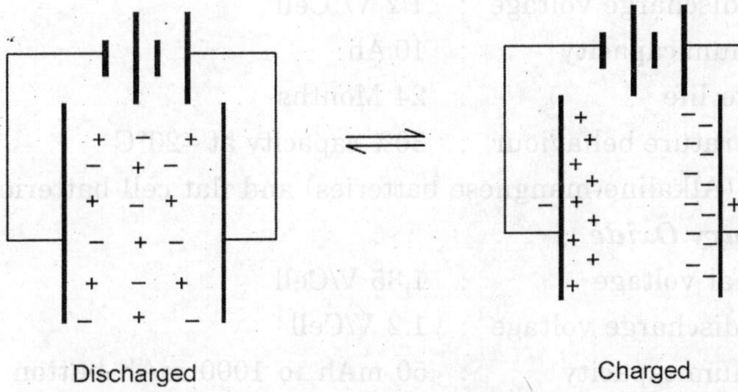

Fig. 5.12: Schematic representation of supercapacitor

The conduction through an ionic liquid is slow, so these capacitors can not be discharged quickly compared with standard capacitors, but can be discharged very quickly compared to batteries.

Typical numbers for capacitors and batteries are given below:

Device	*Volumetric energy density Wh/L*	*Power density W/L*	*Number of charged/ discharged cycles*	*Discharge times*
Batteries	50 – 250	150	$1 - 10^3$	> 1000
Capacitors	0.05 – 5	$10^5 - 10^8$	$10^5 - 10^6$	< 1

Supercapacitors have several advantages over batteries; they can experience virtually indefinite number of cycles, *i.e.*, charging and discharging, they are maintenance free, they work well in high rate discharge, they recharge quickly, and they have no negative environmental impact.

5.6 DESCRIPTION OF AVAILABLE PORTABLE BATTERIES

In brief the description of available portable batteries are as illustrated below:

5.6.1 Primary Batteries

(i) Zn/Carbon (leclanche)

Nominal voltage : 1.5 V/Cell
Mean discharge voltage : 1.2 V/Cell
Maximum capacity : About 7.5 Ah
Storage life : 12 – 24 months
Temperature behaviour : About 50% capacity available at –10°C, inefficient at low temp.

Standardized and flat batteries up to 4 Ah are available.

(ii) Zinc/Carbon (alkaline)

Nominal voltage : 1.5 V/Cell

Mean discharge voltage	:	1.2 V/ Cell
Maximum capacity	:	10 Ah
Storage life	:	24 Months
Temperature behaviour	:	30% capacity at –20°C

Cylindrical cells (Alkaline-mangnese batteries) and flat cell batteries are available.

(iii) Zinc/Mercury Oxide

Nominal voltage	:	1.35 V/Cell
Mean discharge voltage	:	1.2 V/Cell
Maximum capacity	:	50 mAh to 1000 mAh button cells
Storage life	:	24 months and more
Temperature behaviour	:	Applicable from –10°C to +60°C

(iv) Zinc/Silver oxide (valency : 1)

Nominal voltage	:	1.55 V/Cell
Mean discharge voltage	:	1.4 V/Cell
Maximum capacity	:	15 mAh to 180 mAh (Button cells)
Storage life	:	More than 24 months
Temperature behaviour	:	Applicable from –10°C to +60°C

(v) Zinc/Silver oxide (valency : 2)

Nominal voltage	:	1.55 V/Cell
Mean discharge voltage	:	1.2 V/Cell
Capacity	:	30 mAh to 90 mAh (Button cells)
Storage life	:	More than 24 months
Temperature behaviour	:	Applicable from –10°C to +60°C

(vi) Air/zinc (With acidic electrolyte)

Nominal voltage	:	1.45 V/Cell
Mean discharge voltage	:	1.15 V/Cell
Capacity	:	Upto 70 Ah
Storage life	:	More than 12 months
Temperature behaviour	:	Applicable from –10°C to +60°C

(vii) Air/zinc (Alkaline)

Nominal voltage	:	1.4 V/Cell
Mean discharge voltage	:	1.2 V/Cell
Capacity	:	Upto 400 mAh (button cells)
Storage capacity	:	Self discharge less than 1% per year activated only after contact with air
Temperature behaviour	:	Applicable from –10°C to +60°C

(viii) Lithium – $MnO_2/SO_2/CuO_2/SOCl_2$

Nominal voltage	:	1.5 to 3.8 Volts
Capacity	:	Upto about 12 Ah

Storage life	:	Self discharge, less than 1% per year upto 10 years
Temperature behaviour	:	Partly also applicable with extreme temperatures.

5.6.2 Secondary Batteries

(i) Nickel/Cadmium (with mass electrode)

Nominal voltage	:	1.2 V/Cell
Capacity	:	10 mAh to 1000 mAh (button cell)
Storage life	:	Practically unlimited
Temperature behaviour	:	
Discharge	:	–20°C to +45°C
Charging	:	0°C to +45°C

(ii) Nickel/Cadmium (With sintered electrode)

Nominal voltage	:	1.2 V/Cell
Capacity	:	100mAh to 7Ah (cylindrical cell)
		2.4 Ah to 15 Ah (prismatic cell)
Storage life	:	Practically unlimited
Temperature behaviour	:	
Discharge	:	–45°C to +65°C
Charging	:	–20°C to +50°C

(iii) Lead/Lead Oxide

Nominal voltage	:	2.0 V/Cell
Capacity	:	2, 4, 6 and 12 volts units from 1Ah to 63Ah for cylindrical and prismatic batteries
Storage life	:	Recharge after 12 months
Temperature behaviour	:	–30°C to + 50°C

PROBLEMS

1. Write down the characteristics of batteries.
2. What do you mean by Peukert Equation and Ragone plot?
3. Explain in brief about the classification of batteries.
4. Write in details about following:
   (*a*) Zinc-Air cells
   (*b*) Lithium cells
   (*c*) Nithium-Metal Hydride Battery
   (*d*) Reserve (Lithium-Vanadium Pentoxide) battery
   (*e*) Super capacitors
   (*f*) Portable batteries

CHAPTER

6

CORROSION

6.1 INTRODUCTION

One of the most widely accepted definitions of corrosion is that "corrosion is the destruction of metal or alloy (except gold and platinum) by chemical or electro-chemical reaction with its environment". Thus corrosion involves chemical or electro-chemical reaction and the use of the term is restricted to the destruction of metal or alloys. Destruction of wood, for instance, by rotting is not referred to as corrosion of wood. Physical or mechanical wearing away of a metal is not called corrosion but is called erosion. If corrosion and erosion takes place together the destruction of the metal takes place very much faster. The most popular example of corrosion is "rusting of iron".

$$Fe + 2H_2O \longrightarrow Fe(OH)_2$$

$$2Fe(HO)_2 + O_2 \longrightarrow \underset{\text{(Rust)}}{Fe_2O_3.2H_2O}$$

It has been estimated that the money spent in controlling corrosion is of the order of Rs. 50 crores in India, Rs. 250 crores in UK, Rs. 200 crores in Australia and Rs. 400 crores in USA each year. In India due to its tropical climate the corrosion problem is more serious than in the colder countries. Now a days, the metals which are widely used as structural materials are corroded rapidly due to the increasing pollution of water and air.

Following are the main effects of corrosion in industry as well as on industrial equipment:

(*i*) Plant shutdown due to failure

(*ii*) Replacement of corroded equipment

(*iii*) Preventive maintenance such as painting

(*iv*) Loss of efficiency

(*v*) Necessity for overdesign to allow for corrosion

(*vi*) Safety, for example, from a fire hazard or explosion or release of a toxic product or a collapse of construction because of a sudden failure

(*vii*) Contamination or loss of the product *e.g.*, from a corroded container.

(*viii*) Health, for example, from pollution due to a corrosion product or due to an escaping chemical from a corroded equipment.

6.2 CAUSES OF CORROSION

The question arises here is to how and why corrosion occurs? Most of the metals exist in nature *i.e.* in ores and minerals in combined form (compound) as their oxides, carbonates, hydroxy carbonates, sulphides, chlorides and silicates etc., because these compounds represent their thermodynamically stable state. The metals are extracted from these ores. During extraction of metals, considerable amount of energy is required. Consequently, isolated pure metal can be considered in excited state *i.e.*, a higher energy state. Hence, unless the nature of the metal is substantially changed *e.g.*, by alloying, the metal will have a natural tendency to revert back to its natural thermodynamically stable state, *i.e.*, of lower energy state. Hence, when metals are put into use in various forms, they are exposed to environment (such as dry gases, moisture, liquid, etc.), the exposed metal surfaces begin to decay *i.e.*, conversion into more stable compound more or less rapidly.

This is the basic reason for metallic corrosion. However, metals like gold and platinum, that exist in elemental state in nature, naturally have very good corrosion resistance in natural environment.

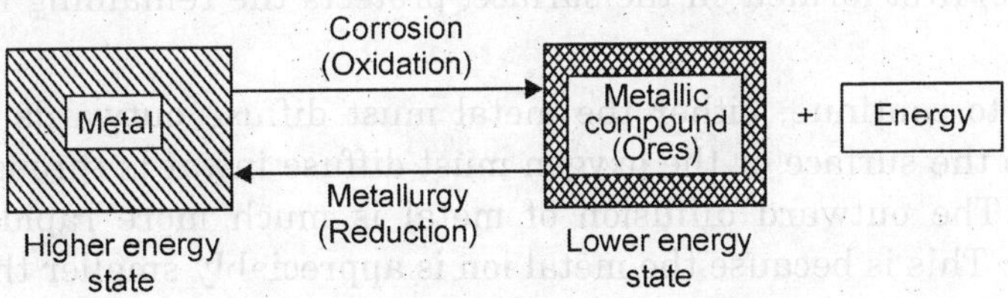

Fig. 6.1 Corrosion and metallurgy process of metals.

As discussed earlier that corroded metal is thermodynamically more stable than pure metal, but due to corrosion, some useful properties of metals such as electrical conductivity, ductility and malleability etc., are lost.

6.3 TYPES OF CORROSION

Corrosion are discussed under the following five categories:

(*i*) Dry corrosion

(*ii*) Wet corrosion

(*iii*) Atmospheric corrosion

(*iv*) Soil corrosion

(*v*) Micro-biological corrosion

6.3.1 Dry Corrosion

Dry corrosion is also known as direct chemical corrosion or direct chemical attack. This type of corrosion occurs mainly through the direct chemical action of environment in the dry state; atmospheric gases such as oxygen, hydrogen, halogen, hydrogen sulphide, sulphur dioxide, nitrogen on anhydrous inorganic liquids with metal surfaces in immediate proximity. Dry corrosion are of three types:

(*i*) Oxidation corrosion

(*ii*) Liquid metal corrosion

(*iii*) Corrosion by other gases

(i) Oxidation Corrosion

One of the most common ways in which metals are attacked by direct action is by interaction with oxygen. Oxidation corrosion is brought about by the action of oxygen at low or high temperatures on metals, usually, in the absence of moisture. Alkali metal (*e.g.* Li, Na, K, Rb, etc.) and alkaline earth metals (Be, Sr, Ca, etc.) are oxidized rapidly at low temperature, whereas almost all metals except Ag, Au and platinum are oxidized at high temperature. Though the formation of an oxide may be a thermodynamically feasible process even at low temperatures the rates of oxidation of the metals in these temperature ranges may be low. However, at higher temperatures the oxidation is also kinetically favoured.

Direct attack of oxygen on metals does not take place to any appreciable extent at atmospheric temperatures. When it does take place a thin film of the solid corrosion product (oxide of the metals) first formed on the surface, protects the remaining metal from further attack.

For oxidation to continue, either the metal must diffuse outwards through the scale (oxide of metals) to the surface or the oxygen must diffuse inwards through the scale to the underlying metal. The outward diffusion of metal is much more rapid than the inward diffusion of oxygen. This is because the metal ion is appreciably smaller than the oxygen ion as well as of much higher mobility. During the oxidation corrosion following chemical reaction take place:

$$2M \longrightarrow 2M^{n+} + 2ne^- \quad \text{...(Loss of electrons)}$$

$$nO_2 + 2ne^- \longrightarrow 2nO^{2-} \quad \text{...(Gain of electrons)}$$

$$2M + nO_2 \longrightarrow 2M^{n+} + 2nO^{2-}$$

where M represents metal

M^{n+} represents metal ions

nO^{2-} represents oxide ions

O_2 represents oxygen and

$[M^{n+} + nO^{2-}]$ represents metal oxide.

The above statements regarding oxidation corrosion can be easily understood using Fig. 6.2.

When oxidation occurs, a thin layer of oxide is formed on the surface of metal. A layer is called film when its thickness is less than 300Å and is called as scale when its thickness is more than 300Å. The layer of oxide formed on the metal surface are of the following nature:

(i) Stable

(ii) Unstable

(iii) Volatile

(iv) Porous.

(i) The stable layer (film) behaves as protective coating in nature and thus shielding the metal surface. Hence, such a film can cut off penetration of attacking oxygen to the underlying metal, *e.g.*, oxide film of Cu, Al, Pb, Sn, Pt, etc. acts as a protective coating and thus further corrosion is prevented.

(ii) When unstable oxide film is formed, the decomposition of metal oxide into metal and oxygen takes place. Hence, the metals further do not go under oxidation corrosion *e.g.*, Au, Ag and Pt do not undergo oxidation corrosion.

(iii) The moment at which oxide film is formed, it gets volatilized, thereby leaving the underlying metal surface exposed for further attack, leading to continuous and rapid corrosion, *e.g.*, molybdenum oxide is volatile.

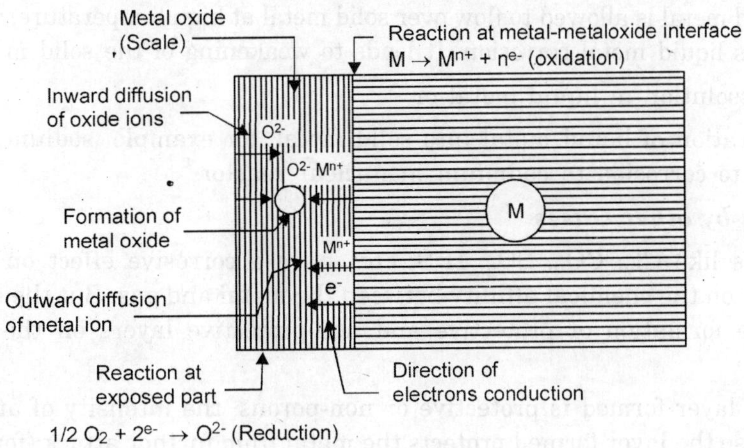

Fig. 6.2: Schematic representation of oxidation mechanism of metal.

$$2Mo + 3O_2 \longrightarrow \underset{\text{(volatile)}}{2MoO_3}$$

(iv) The oxide has less volume, the deposit is porous or crack and open space is left through which atmospheric oxygen diffuses to the metal surface. Thus, corrosion proceeds till the entire metal is converted into metal oxide.

Pilling-bed worth rule: According to this rule, we can define an oxide is protective (non-porous) or non-protective by means of the volume of the oxide. Therefore, "an oxide is protective, if the volume of the oxide is at least as great as the volume of the metal from which it is formed". On the other hand, "if the volume of the oxide is less than the volume of metal from which it is formed the oxide layer is porous and hence non-protective, because it cannot prevent the approach of oxygen to the fresh metal surface below". For example, alkali and alkaline earth metals (Li, Na, K, Rb, Be, Ca, Sr, etc.) on oxidation produce oxide deposits of smaller volume than the respective metals from which they were formed. This results in the formation of a porous layer through which oxygen can diffuse to bring about further attack on the metal, thus corrosion continues non-stop. On the other hand, heavy metals like Al form oxides of greater volume than the metal from which they were produced. These non-porous continuous oxide films prevent the diffusion of oxygen and hence the role of further attack decreases with increase in the thickness of the oxide film.

However, as the film grows thicker, their tendency to crack and scale out becomes greater and particularly at high temperatures, intermittent attack of metal may take place at the exposed parts of the metal (due to cracking of the oxide film).

The ratio of the volume of the oxide to the volume of the metal consumed is known as Pilling bed worth ratio.

It is expressed as:

$$\text{Pilling-bed worth ratio} = \frac{\text{Volume of the oxide formed}}{\text{Volume of the metal consumed}}$$

(ii) Liquid Metal Corrosion

When liquid metal is allowed to flow over solid metal at high temperature, corrosion occurs, and is known as liquid metal corrosion. It leads to weakening of the solid metal due to:

(*i*) its dissolution in liquid metal or

(*ii*) penetration of liquid metal into solid metal, for example, sodium metal (coolant) leads to corrosion to cadmium in nuclear reactor.

(iii) Corrosion by other Gases

Some gases like Cl_2, CO_2, SO_2, H_2S, etc., have a corrosive effect on a metal which depends mainly on the chemical affinity between the metal and gas. But the degree of attack depends on the formation of protective and non-protective layers on the metal surface. For example,

- If the layer formed is protective or non-porous, the intensity of attack decreases, because the layer formed protects the metal from further attack (formation of AgCl layer; attack of Cl_2 on Ag).
- If the layer formed is non protective or porous, the surface of the whole metal is gradually destroyed (formation of $SnCl_4$ volatile layer; attack of dry Cl_2 on Sn).

6.3.2 Wet Corrosion

Wet corrosion is sometimes also known as indirect corrosion or electro-chemical corrosion or immersed corrosion. This type of corrosion occurs mostly under wet or moist conditions through the electrochemical reaction, *i.e.* through the formation of short circuited galvanic cells. This type of corrosion occurs mainly under the following conditions:

(*i*) Where a conducting liquid is in contact with metal or

(*ii*) Where two dissimilar metals or alloys are either immersed or dipped partially in a solution.

The modern electro-chemical theory of corrosion is based on Nernst theory. According to this theory, the corrosion of a metal in aqueous solution may be thought of as a two step process, *i.e* oxidation and reduction. This is due to the existence of separate 'anodic' and 'cathodic' areas/parts between which current flows through the conducting solution.

At anodic area, oxidation reaction, *i.e.*, liberation of free electrons takes place, so anodic metal is destroyed by either dissolving or assuming combined state such as oxide, etc. Therefore, corrosion always occurs at anodic areas.

$$\underset{M}{\text{Metal}} \xrightarrow{\text{Oxidation}} \underset{M^{n+}}{\text{Metal ion}} \left\{ \begin{array}{l} \rightarrow \text{Dissolved in solution} \\ \rightarrow \text{Forms their oxide} \end{array} \right.$$

Whereas, at cathodic areas, reduction reaction, *i.e.* gain of electrons takes place. At cathodic part dissolved constituents in the conducting medium accept the electrons to form some ions like OH^-, O^{2-} etc.

The metallic ions formed at anodic part and non metallic ions formed at cathodic part get diffused towards each other through conducting medium and form a corrosion product somewhere between anode and cathode.

The electrons set free at the anode flow through the metal and are finally consumed in the cathodic reaction. Thus, overall electro-chemical corrosion involves:

(*i*) Formation of anodic and cathodic areas/parts in contact with each other.

(*ii*) Presence of a conducting medium

(*iii*) Corrosion of anodic areas only and

(*iv*) Formation of corrosion product somewhere between anodic and cathodic areas.

Wet corrosion involves flow of electron current between the anodic and cathodic areas. The anodic reaction involves in dissolution of metal as corresponding metallic ions with the liberation of free electrons.

$$\underset{\text{metal}}{M} \xrightarrow{\text{Oxidation}} \underset{\text{metal ion}}{M^{n+}} + \underset{\text{electrons}}{ne^-}$$

Whereas, the cathodic reactions accept electrons by means of two ways; evolution of hydrogen or by absorption of oxygen. The evolution of hydrogen and absorption of oxygen in cathodic reaction depend on the nature of corrosive environment.

Evolution of Hydrogen: The process of corrosion in which H_2 is liberated is called evolution of hydrogen type corrosion. This type of corrosion usually takes place in acidic environment. Thus, in acidic medium (absence of oxygen) hydrogen ions acquire electrons with the liberation of H_2 gas in cathodic reaction, whereas at anodic reaction metal atoms lose their electrons to the environment and pass into solution in the form of +ve ions.

$$2H_3O^+ + 2e^- \longrightarrow 2H_2O + H_2 \uparrow \quad \text{(cathodic reaction)}$$

In neutral and alkaline medium (in the absence of oxygen) the reaction taking place will be

$$2H_2O + 2e^- \longrightarrow 2OH^- + H_2 \uparrow \text{ (cathodic reaction)}$$

whereas

$$Fe \longrightarrow Fe^{2+} + 2e^- \text{ (anodic reaction)}$$

e.g., if iron metal is used, the dissolution of iron as Fe^{2+} ions with the evolution of electrons in anodic area (reaction),

$$Fe \xrightarrow{\text{Oxidation}} Fe^{2+} + 2e^-$$

These electrons flow through the metal, from anode to cathode, where H^+ ions of acidic solution are eliminated as H_2.

$$2H^+ + 2e^- \xrightarrow{\text{Reduction}} H_2$$

The overall reaction will be

$$Fe + 2H^+ \longrightarrow Fe^{2+} + H_2$$

Therefore, this type of corrosion causes displacement of hydrogen ion from the acidic solution by metal ions.

The above statements can be easily understood by Fig. 6.3.

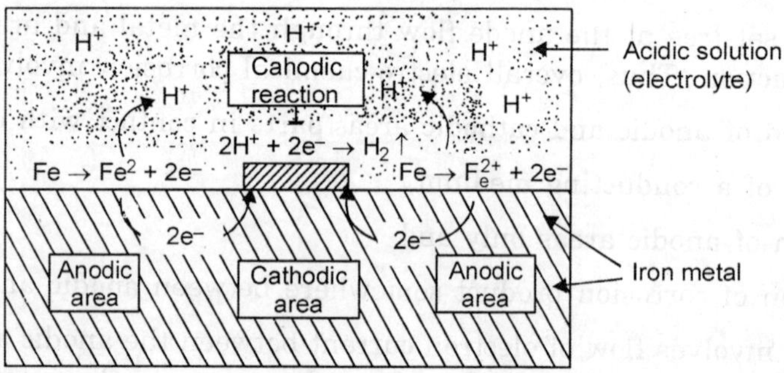

Fig. 6.3 Diagrammatic representation of mechanism of wet corrosion by hydrogen evolution.

Thus, from the above diagram, it is clear that acidic areas are very large in comparison to cathodic areas. All the metals, above hydrogen in the electro-chemical series have a tendency to get dissolved in acidic solution with simultaneous liberation of hydrogen.

Absorption of Oxygen: This type of corrosion occurs in neutral environment (such as NaCl solution used as electrolyte). The popular examples of absorption of oxygen type corrosion is, rusting of iron in neutral aqueous solution of NaCl in the presence of oxygen (atmospheric).

The surface of iron usually covered with a film of iron oxide. If there are some cracks in iron oxide film, an acidic area is developed on the surface, whereas the wet metal parts act as cathodic area. In this corrosion reaction cathodic area is much more large in comparison to anodic area.

The above, overall reaction mechanism can be represented diagrammatically in Fig. 6.4.

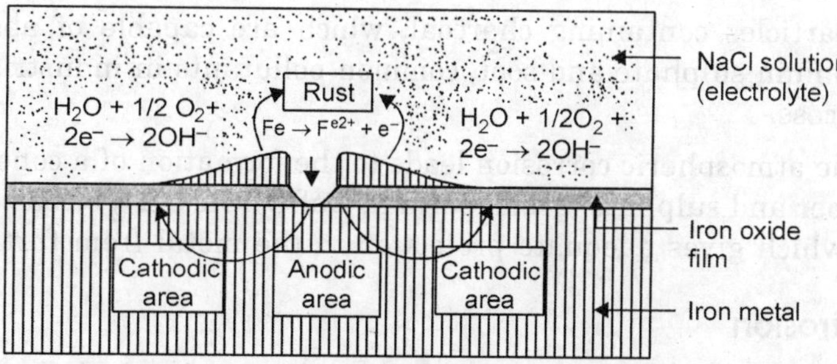

Fig. 6.4 Diagrammatic representation of wet corrosion mechanism by absorption of oxygen type corrosion.

Thus, from the above diagram it is clear that at the anodic area, iron metal dissolves with the liberation of electrons as

$$Fe \xrightarrow{\text{Oxidation}} Fe^{2+} + 2e^-$$

The above evolved electrons flow from anodic to cathodic areas through iron metal where electrons are taken up by the dissolved oxygen as

$$2e^- + O_2 + H_2O \xrightarrow{\text{Reduction}} 2OH^-$$

The OH^- ions formed at cathode as well as Fe^{2+} ions formed at anode combine and form the $Fe(OH)_2$ as precipitate

$$Fe^{2+} + 2OH^- \longrightarrow Fe(OH)_2 \downarrow$$

In the presence of sufficient oxygen, $Fe(OH)_2$ can be easily oxidized into ferric hydroxide $[Fe(OH)_3]$

$$4Fe(OH)_2 + 2H_2O + O_2 \longrightarrow \underset{\text{yellow rust}}{Fe(OH)_3} \downarrow$$

actually $Fe(OH)_3$ corresponds to $Fe_2O_3 . H_2O$.

If the supply of oxygen is limited then black anhydrous magnetite *i.e.*, ferrosoferic oxide is formed as

$$6Fe(OH)_2 \longrightarrow Fe_3O_4(Fe_2O_3 . FeO) . 6H_2O$$

As we know that Fe^{2+} ion is smaller in size than OH^- ion, the Fe^{+2} ion will diffuse more rapidly, so corrosion occurs at the anode, but rust is deposited at the cathode or near the cathode.

6.3.3 Atmospheric Corrosion

Atmospheric corrosion is the result of conjoint action of two factors, viz., oxygen and moisture. If one of these factors is missing, corrosion does not occur, *e.g.*, if iron is exposed to an atmosphere containing humidity more than 60% *i.e.*, critical humidity, rusting of iron starts and increases rapidly with increasing relative humidity. Even at high relative humidities corrosion appears to be negligible as long as the atmosphere is dust free. Steel, also does not rust if covered with ice. This is only due to the lack of water as the electrochemical attack requires water in the form of liquid. Atmospheric corrosion also increases if air is polluted by smoke or gases like SO_2 and vapours in marine environment.

Some dust particles containing charcoal, which are capable of absorbing SO_2, solid particles of ammonium sulphate and soot, common pollutants in industrial areas accelerate the corrosion process.

On copper, the atmospheric corrosion leads to the formation of a coherent and adherent layer of basic copper and sulphate (green patina), $CuSO_4.Cu(OH)_2$ or basic copper carbonate $CuCO_3.Cu(OH)_2$ which gives adequate protection to the metal from further attack.

10.3.4 Soil Corrosion

The various factors that are responsible for soil corrosion include:

(*i*) Acidity of the soil

(*ii*) Moisture content

(*iii*) Content of electrolyte

(*iv*) Micro-organisms present

(*v*) Content of organic matter and

(*vi*) Physical properties of the soil.

In non-acid soils, the conductivity which depends on moisture and electrolyte content, is the major factor governing the corrosive character. In such soils, corrosion takes place due to the formation of differential aeration couples and the rate of corrosion mainly depends upon the resistance between the anodic and cathodic areas and the rate of the arrival of the oxygen to the cathodic areas.

Thus, when a cable or pipe passes under a paving, the portion under the paving has less access to oxygen than the one lying under unpaved soil. Hence, the portion under the paving becomes anodic and suffers corrosion.

In highly acidic soils, the corrosion taking place is of hydrogen type with anodic and cathodic areas being very close to each other. The conductivity of the soil is not of much importance in this case.

If the soils contain large amount of organic matter, the formation of soluble metal complexes and the peptization of corrosion products may accelerate the corrosion process as compared to what is expected in the soils relatively free from organic matter.

Soils rich in gravel and sand are more porous and hence are generally more aerated and the corrosion of a metal pipe under such soils depends mainly on the moisture and salt content.

Water-logged soil may generate anaerobic bacteria which may create conditions for microbiological corrosion.

Buried pipelines or cables passing from one type of soil to another, *i.e.*, from less aerated soil to more aerated soil may get corroded due to differential aeration. Other corrosion factors in this case are the differences in pH and the presence of unburnt carbon in the cinders.

Air pockets in the soil, if present, may also cause corrosion due to differential aeration.

6.3.5 Microbiological Corrosion

Corrosion caused by the metabolic activity of various micro-organisms is called microbiological corrosion. The microorganism can develop and propagate in an environment in the presence of oxygen, sometimes in the absence of oxygen also. Therefore, they are termed as aerobic and anaerobic.

Following three types of microorganism are responsible for microbiological corrosion.

(*a*) Aerobic Microorganism

(*i*) Sulphur bacteria

(*ii*) Iron and manganese bacteria

(*b*) Anaerobic microorganism

(*i*) Sulphate reducing bacteria

(*c*) Film forming microorganisms

Sulphur bacteria, the thioracillus oxidize sulphur present in their cells to yield sulphuric acid, which attacks the metal like iron. Such types of bacteria grow in acidic conditions say pH from 0 to 1.

Iron and manganese microorganisms are aerobic bacteria and live by taking into their cells iron and manganese ions, which they digest in the presence of oxygen forming insoluble hydrates of iron and manganese dioxide, which are then thrown out of their bodies. Iron bacteria grow in stagnant or running water at 5 to 40°C and pH between 4 to 10.

Sulphur reducing bacteria, the Sporovobrio desulphuricous are responsible for anaerobic corrosion of iron and steel. They require sulphur and oxygen for their propagation and grow at 25 to 30°C and pH between 5 to 9.

Bacteria, fungi, algae and diatoms are film forming microorganisms. They can form microbiological film on an iron surface. Such films are capable of maintaining concentration gradients of dissolved salts, acids and gases on the iron surface, thereby leading to the formation of local biological concentration cells and consequent corrosion.

6.4 THEORIES OF CORROSION

In atmospheric corrosion the metals tend to revert back to the state in which they occur in nature i.e. the ores, because that is the most stable state. Such compounds always have less free energy than the pure metal. Fundamentally this is the cause of corrosion. Several theories of corrosion have been proposed from time to time. Following are the main theories which are mainly applicable to corrosion:

(*i*) The acid theory

(*ii*) Chemical attack theory and

(*iii*) Electrochemical theory.

6.4.1 The Acid Theory

In the early nineteenth century it was demonstrated that for rusting of iron the presence of both oxygen and water is necessary. It was also noticed that rusting takes place faster in acid solutions than in neutral or alkaline solutions. This was further confirmed by the detection of ferrous and ferric carbonates (due to carbonic acid in water) in iron rust which mostly consists of hydrated ferric oxide. This led to the acid theory of corrosion which considers acid to be the main cause of corrosion.

$$Fe + O + 2CO_2 + H_2O \longrightarrow Fe\ (HCO_3)_2$$

$$2Fe\ (HCO_3)_2 + H_2O + O \longrightarrow 2Fe(OH)\ CO_3 + 2CO_2 + 2H_2O$$

$$2Fe\ (OH)\ CO_3 + 2H_2O \longrightarrow 2Fe\ (OH)_3 + 2CO_2$$

This theory is supported by the fact that:

(*i*) the presence of ferrous and ferric carbonates and hydrated ferric oxide, and

(*ii*) retardation of rusting in presence of lime or NaOH.

6.4.2 Chemical Attack Theory or Direct Chemical Attack Theory

This theory explains the so-called chemical or dry corrosion. Direct chemical attack by dry gases on a metal at atmospheric temperature does not take place. However, whenever corrosion takes place by direct chemical attack, a solid film of the corrosion product is usually formed on the surface of the metal which protects the metal from further corrosion. However, when a soluble or liquid corrosion product is formed, then the metal is further exposed to attack.

6.4.3 Electrochemical Theory

It is now accepted that most of the common cases of corrosion can best be explained on the basis of the electrochemical theory. At ordinary temperatures the corrosion process involves oxidation. Very few cases of non-oxidative corrosion are met with under common conditions. Oxidation of a metal involves its loosing of electrons, thereby acquiring a positive charge. In the solid state the electrons are carried away through a metal conductor and, in solution, as ions.

If a piece of metal is dipped in water or a solution, it has a tendency to go into solution. Some of its atoms loose electrons and go into solution as positively charged ions. These ions are held near the metal piece because of electro-static attraction and the process stops unless this balance is disturbed.

Therefore, for the progress of corrosion anodic and cathodic areas must exit in electrical contact with each other. Thus, whenever corrosion is in progress a number of tiny galvanic cells are active.

Galvanic cells: On a corroding surface galvanic cells may be formed in several ways. The principle of their working can be explained by considering a Daniell cell which consists of zinc and copper plates dipping in solutions having Z^{++} ion and Cu^{++} ions, the solutions being separated by a porous partition. The plates are connected by a metallic conductor. The e.m.f. developed is due to two separate reactions taking place at the electrodes. Zinc atoms have a tendency to go into solution as Zn^{++} making two electrons available. This is an oxidation reaction:

$$Zn\ (s) \longrightarrow Zn^{++}\ (aq) + 2e^-$$

The electrons move along the conductor and are responsible for the reduction reaction at the copper plate:

$$Cu^{++}\,(aq) + 2e^- \longrightarrow Cu\,(s)$$

Thus, Zn dissolves and Cu is deposited at its cost. Exactly similar reaction occurs in corrosion.

In this cell the Zn plate is called an anode because oxidation reaction, *i.e.*, liberation of electrons, takes place at the zinc plate. Similarly, because reduction reaction *i.e.*, consumption of electrons, takes place at the copper plate, the plate is called a cathode. The greater is the e.m.f., the faster will be the corrosion.

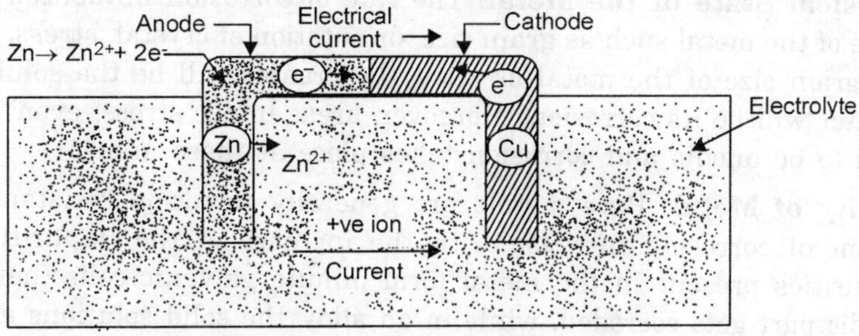

Fig. 6.5 Diagrammatic representation of galvanic corrosion.

When two dissimilar metals like Zn and Cu are electrically connected and exposed to an electrolyte, the metal higher in electrochemical series undergoes corrosion. This type of corrosion is called "galvanic corrosion". In the above example, Zn, higher in electrochemical series forms the anode and is attacked and gets dissolved; whereas Cu, lower in electrochemical series acts as cathode.

The details about electrochemical corrosion have already been discussed under the heading.

6.5 FACTORS INFLUENCING CORROSION

Following are the two main factors influencing the rate and extent of corrosion:

(*a*) Nature of the metal

- (*i*) Physical state of the metal
- (*ii*) Purity of metal
- (*iii*) Position in the galvanic series
- (*iv*) Nature of the oxide films
- (*v*) Volatility of corrosion product
- (*vi*) Solubilities of the product of corrosion
- (*vii*) Relative areas of the anode and cathode

(*b*) Nature of the environment

- (*i*) Temperature
- (*ii*) Humidity

(*iii*) pH

(*iv*) Nature of electrolyte

(*v*) Conductance of corroding medium

(*vi*) Presence of impurities in atmosphere

(*vii*) Presence of suspended particles in atmosphere

(*viii*) Formation of oxygen concentration cell.

6.5.1 Nature of the Metal

(*i*) **Physical State of the Metal:** The rate of corrosion influenced by the physical state of the metal such as grain size, orientation of crystal, stress, etc. The smaller the grain size of the metal or alloy, the greater will be the solubility and hence greater will be its corrosion. Moreover, areas under stress, even in a pure metal, tend to be anodic and corrosion takes place at these areas.

(*ii*) **Purity of Metal:** Pure metals are generally corrosion resistant. The rate and extent of corrosion increase with the increase of the impurities, because the impurities present in the metal form minute galvanic cells (heterogeneity) and anodic part gets corroded, while in an alloy, the solid solutions are homogeneous and do not form galvanic cells.

Lesser is the percentage purity of a metal, faster is the rate of corrosion, *e.g.*, in the case of zinc metal

% Purity	99.999	99.99	99.95	99.0
Corrosion Rate	1	2650	5000	7,200

(*iii*) **Position in the Galvanic Series:** The extent of corrosion depends upon the position of the metal in the electrochemical series and galvanic series. When two metals are in electrical contact in the presence of an electrolyte, the more active metal (or higher up in the series) suffers corrosion. The rate and severity of corrosion depends upon the difference in their positions, and greater is the difference, the faster is the corrosion of the anodic metal or alloy. The electrochemical series and galvanic series of metal are given in the Tables 6.1 and 6.2 respectively

(*iv*) **Nature of the Oxide Films:** In aerated atmosphere practically all metals get covered with a thin surface film of metal oxide having a thickness of a few angstroms ($1Å = 10^{-10}$ m). The ratio of the volume of the metal oxide to the metal is known as specific volume ratio. Greater the specific volume ratio, lesser is the oxidation corrosion rate *e.g.*, the specific volume ratio of Ni, Cr and W are 1.6, 2.0 and 3.6 respectively, which indicates that the rate of oxidation at elevated temperature is least for tungsten (W).

(*v*) **Volatility of Corrosion Product:** If the corrosion volume is more volatile, then metal surface is exposed for further attack, *i.e.*, corrosion rate will be more, *e.g.*, Molybdenum oxide (MoO_3), the oxidation corrosion product of molybdenum is volatile.

(vi) **Solubilities of the Product of Corrosion:** If the corrosion product is soluble, corrosion of the metal will proceed faster. On the other hand, if corrosion product formed is insoluble, then the protective film formed tends to suppress corrosion.

(vii) **Relative Areas of the Anode and Cathode:** When two steel plates of the same area are separately connected, one to a copper plate (cathode) having the same area and the other to a copper plate having a much larger areas, it can be observed that the latter couple produces greater amount of current and the corrosion of the steel anode in this case will be much greater than the first one. It may be broadly concluded that when anodic polarization is negligible and conductance remaining more or less same, corrosion of the anode is directly proportional to the area of the cathode.

Table 6.1: Standard electrode potential (at 25°C)

Metal-ion Electrode Reaction	*Standard Electrode Potential (E_o) in volts*	
Li/Li^+	–3.045	(Reactive metals) ↑
K/K^+	–2.92	
Ca/Ca^{++}	–2.84	
Na/Na^+	–2.71	
Mg/Mg^{++}	–2.37	
Al/Al^{+++}	–1.67	
Mn/Mn^{++}	–1.05	
Zn/Zn^+	–0.762	
Cr/Cr^{+++}	–0.71	
Fe/Fe^{++}	–0.441	
Cd/Cd^{++}	–0.402	
Co/Co^{++}	–0.277	
Ni/Ni^{++}	–0.25	
Sn/Sn^{++}	–0.136	
Pb/Pb^{++}	–0.126	
$H_2/2H^+$	0.000	(Reference)
Cu/Cu^{++}	0.345	
Cu/Cu^+	0.522	
Hg/Hg^+ or $Hg_2/2H^+$	0.799	
Ag/Ag^+	0.800	
Hg/Hg^{++}	0.854	
Pd/Pd^{++}	0.987	
Pt/Pt^{++}	1.2	↓
Au/Au^{+++}	1.42	(Noble metals)

Table 6.2: Galvanic series (on the basis of relative oxidation potential in sea water)

More Anodic or Active	*(or Corroded end)*
Magnesium Magnesium alloys Zinc Aluminium Aluminium alloys Low carbon steel Cast iron Stainless steel (active) Lead-tin alloys Lead Tin Brass Copper Bronze Copper-nickel alloys Inconel Silver Stainless steel (passive) Monel Graphite Titanium Gold Platinum	↑ Active (anodic) Noble (cathodic) ↓
More cathodic or inactive	

6.5.2 Nature of the Environment

(*i*) **Temperature:** With increase of temperature of environment the reaction as well as diffusion rate increase, thereby corrosion rate is generally increased.

(*ii*) **Humidity:** The higher the moisture content in the atmosphere, the greater is the rate and extent of the corrosion because moisture acts as medium for oxygen in air and behaves as an electrolyte in air e.g. atmospheric corrosion of iron is slow in dry air compared to moist air.

(*iii*) **pH:** Acidic media are generally more corrosive than alkaline and neutral media. The amphoteric metals such as Al, Zd, Pb, etc. dissolve corroded in alkaline solution. The corrosion rate of iron in oxygen free water is slow, until pH is below 5. The corresponding corrosion rate in presence of oxygen is much higher. Zinc is rapidly corroded, even in weakly acidic solutions such as carbonic acid suffers minimum corrosion at pH = 11.

(*iv*) **Nature of electrolyte:** Chloride ions present in the medium destroy the passive film and corrode many metals and alloys. On the contrary, some anions like silicate may form an insoluble product (silica gel) which inhibit corrosion.

(*v*) **Conductance of corrosion medium:** The corrosion current depends upon the internal resistance of the short circuited cells which in turn depends on the conductance of the electrolytic medium. If the conductance of the soil is more, the corrosion of underground structures will also be more, consequently there will be more severe damage to the metallic structures in the presence of stray currents from power leakage. Conductance of clay and mineral soil is more than that of dry sandy soil, hence corrosion of underground metallic structures is more in clay and mineralized soil.

(*vi*) **Presence of impurities in atmosphere:** Atmosphere, in the vicinity of industrial areas, contains corrosive gases like SO_2, H_2S and CO_2 and fumes of HCl and H_2SO_4 etc. In the presence of these gases, the acidity of the liquid, adjacent to the metal surfaces, increases and its electrical conductivity also increases. This consequent results in an increase of corrosion on the exposed metal surfaces. Similarly, in marine atmosphere, presence of Na^+ and Cl^- leads to increase the conductivity of liquid layer in contact with the metal surfaces thereby increases the corrosion.

(*vii*) **Presence of suspended particles in atmosphere:** Two types of suspended particles are present in atmosphere viz., chemically active and chemically inactive. The chemically suspended particles like NaCl, $(NH_4)_2SO_4$ absorb moisture and thus act as strong electrolytes, thereby enhance corrosion, whereas chemically inactive suspended particles like charcoal, absorb both moisture as well as sulphur gases and thus slowly enhances corrosion rate.

(*viii*) **Formation of Oxygen Concentration Cell:** The rate of corrosion increases with increasing supply of oxygen. Hydrogen depolarisation also increases under these conditions. Further differential aeration will also promote corrosion by the formation of concentration cells. The region where the oxygen concentration is less becomes anodic and suffers corrosion. The mechanism of corrosion by differential aeration has already been discussed under the electrochemical theory of corrosion.

6.6 PROTECTION FORM CORROSION

Protection against corrosion means not allowing corrosion reaction to take place. In most practical cases the cost of a metal or alloy of fabrication is such that the most resistant material cannot be used and some steps have to be taken to prevent corrosion, *e.g.* noble metals do not corrode but they cannot be used for common purposes. In some cases this can be done directly, *e.g.* by painting and in some cases indirectly, *e.g.* by attaching blocks of magnesium to underground pipelines.

The method used depends largely on the environment to which the metal is exposed. The metal should be cleaned or 'descaled' before protective measures are adopted. This pretreatment usually consists of 'degreasing' and descaling. Oily and greasy surface films are dissolved out by exposure to volatile organic solvents such as trichloroethylene ($CHCl{=}CCl_2$), tetrachloroethylene ($Cl_2C{=}CCl_2$), etc. Descaling can be effected either mechanically or chemically.

Since water and oxygen are essential for rusting, the best way would appear to be to protect the metal from these two. The various measures taken for protection against corrosion can be studied under:

(*i*) Design and material selection

(*ii*) Cathodic and anodic protection, and

(*iii*) Protective coating

6.6.1 Passivity

It is also known as passivation. It can be defined as a phenomenon in which metal or an alloy exhibits a much higher corrosion resistance than its position in the electrochemical series would indicate. This is due to the formation of a thin (0.0004 mm thick) but highly protective film of oxide on the surface of the metal or alloy. The passivated metal can be rendered active by a change in the environmental conditions.

Metals which exhibit passivity include iron, chromium, nickel, titanium and alloys of these metals. Passivation is usually associated with oxidising environment and are characterised by the formation of a very thin protective oxide film. However, it should be emphasized that a metal which is passive in one environment may not necessarily be so in a different environment, because passivity results from a continuing reaction between a metal and its environment. When a metal is passive, it corrodes at a very slow rate. Hence, passivity offers an excellent method of corrosion control.

The passivation is not static in nature but quite stable under certain environmental conditions, says oxidizing conditions, e.g. the metals or alloys of Al, Cr, Ti, etc. exhibits high corrosion resistance in oxidizing environments but are quite unstable in reducing environments. The passivity of some metals are in the order of:

$$\text{Ti} \rightarrow \text{Al} \rightarrow \text{Cr} \rightarrow \text{Be} \rightarrow \text{Mo} \rightarrow \text{Mg} \rightarrow \text{Ni} \rightarrow \text{Co} \rightarrow$$
$$\text{Fe} \rightarrow \text{Mn} \rightarrow \text{Zn} \rightarrow \text{Cd} \rightarrow \text{Sn} \rightarrow \text{Pb} \rightarrow \text{Cu}.$$

The presence of oxide films may affect the initial rate of attack by acids. Iron dissolves readily in very dil. HNO_3, but as the concentration of the acid is raised, its oxidizing power rises and a point is reached at which the acid directly oxidizes the metal to its oxide. The layer of the oxide formed on the surface makes the iron passive. The passive iron has a high, positive electrode potential and is not easily corroded. The oxidized film is easily damaged by scratching, heating or because of the presence of chloride ions. Under such circumstances, the metal continues to corrode. However, some passive films are not so easily damaged as that of iron, and even if there are cracks in film, these films can readily heal themselves in an oxidising atmosphere stainless steel composed of 18% Cr and 8% Ni is a good example of this. The highly protective coating of Cr_2O_3 is believed to be responsible for the corrosion resistance. Other example of increased resistance to oxidation by alloying are silicon containing cast iron and silicon containing molybdenum.

6.6.2 Corrosion Inhibitors

Corrosion inhibitors are the substances which retard or stop the corrosion reaction. They may be inorganic or organic substances. Some of these are polarisers and act by retarding cathodic or anodic reactions. Their exact action is varied, *e.g.* formation of an insoluble salt or oxide on the metal surface, adsorption on the metal surface, etc.

6.6.2.1 Inorganic Inhibitors

Inorganic inhibitors such as silicates, chromates, borates, etc. suppress the rate of corrosion by acting on the anode. Alkaline sodium nitrite alone or in combination with other inhibitors such as phosphate have been used to control the corrosion of tankers and pipelines. Sodium benzoate has been used as an inhibitor for mild steel and in preventing corrosion in cooling system such as automobile radiators. A mixture containing 0.1% $NaNO_2$ and 1.5% sodium benzoate has been used as inhibitors in antifreeze solutions. Lime acts as a cathodic inhibitor by precipitating $CaCO_3$ in water containing temporary hardness or dissolved CO_2. Colloidal particles of $CaCO_3$ having +ve charge are attracted to the cathodic areas and get deposited there, thus, reducing the corrosion current.

6.6.2.2 Organic Inhibitors

Organic inhibitors act by different mechanisms *e.g.*, organic colloidal inhibitors form protective layers on the metal surface by adsorption. Surface active reagents known as surfactant, containing polar groups promote spreading and oriented adhesion to the surface, thus, forming a protective film. Organic bases *e.g.*, amines, pyridines, quinoline and their derivatives contain hydrophobic group or radicals. These positively charged cationic groups attack themselves to the cathodic areas through the nitrogen and provide inhibition. The efficiency of inhibition depends upon size and number of the alkyl groups, *e.g.*, primary amyl amine ($C_5H_{11}NH_2$) has been found to be more effective as inhibitor than primary ethyl amine ($C_2H_5NH_2$).

High molecular weight amines derived from rosin have been used as corrosion inhibitors. The product obtained by rosin amine and pentachlorophenol was used in castings for underground pipelines to protect them from corrosion by soil bacteria. Organic inhibitors such as soluble salts of rosin *e.g.*, stearate or naphthenate have been used as inhibitors in the various metal cleaning operations like acid "pickling" of metals, cleaning of boilers, condensers, heat exchangers, chemical equipment, pipelines, etc, to prevent undue and excess corrosion of the metal while removing rust, scales and deposits.

Some vapour phase inhibitors (VPI) such as dicyclohexylammonium nitrite and cyclohexylamine carbonate are used to inhibit the corrosion in storing air craft engines as well as during storage, packing, shipping, etc. octadecyl amine, hexadecyl amine and dioctadecyl amine do not come under VPI but they provide protective action by forming impervious non-wettable surface films. These are used in radiator compounds and boiler waters. Amine salts, morpholine and sodium benzoate are particularly useful for impregnation of packing materials.

It should be emphasized here that an inhibitor should be used with caution. Improper and insufficient use of an inhibitor may even accelerate corrosion, *e.g.*, addition of chromate as inhibitor in the presence of hydrogen polarization may have depolarization effect and corrosion may be accelerated, while insufficient use of an inhibitor may not provide the protective film over the entire anode and the small areas left exposed, suffer intense corrosion leading to pitting.

6.6.3 Protective Coatings

The metal surface may be coated with metallic or non-metallic coatings for protection against corrosion. Most of these coatings act by cutting off contact between the metal and the corrosive environment. The relative merits of the metals used for such protective purposes depends on the reactivity of the metal, relative to the metal to be protected, and also on the ability of the protecting metal to form a self protecting layer. Metals such as Zn, Ni, Sn, Cd and Cr which are capable of self protection through their respective oxides, are used for protecting iron from rusting. Cd, Ni and Sn, having more positive standard electrode potentials, would be cathodic to Fe, whereas Zn and Cr would be anodic to Fe. So iron would be expected to be corrode when in contact with the more noble Cd, Ni and Sn, and not to corrode in contact with Zn and Cr.

Coatings which are applied to the surface of metals protect these either:

(*i*) By preventing air and moisture from reaching the surface of the metal to be protected, or

(*ii*) By becoming the anode portion of a corrosion cell.

Before any coating can be applied it is necessary to properly prepare the metal surface.

Surface Preparation

If the metal surface has not been properly prepared, if it is not clean, the protective coating will not stick to the surface and not only will the purpose be defeated but more severe corrosion may result at points exposed to the corrosive environment.

The surface imperfection, besides physical defects is due to the presence of two kinds of substances: Oil and grease and rust or scale. Oil and grease are removed by washing with soap or alkali solution in water. Several patented compositions containing alkalis, phosphates and silicates are marketed for removing oil. The use of solvents mostly chlorinated hydrocarbons, has become quite widespread in the U.S.A.

Rust or scale is mostly removed by dipping the material in dil. (5 to 10%) H_2SO_4 and by mechanical rubbing with a wire brush. Large equipment or bridges are cleaned by the use of wire brush or by sand blasting. Another method of cleaning is by applying flame from a kerosene burner. The old paint, oil or grease burns and the scale is removed because of sudden expansion of the metal.

A surface may also be cleaned by making it a cathode and passing high density current when the hydrogen gas evolved dose the cleaning.

It the surface of metal is not smooth it is made smooth by mechanical polishing or by electroplating. The clean surface catches dust and gets corroded quickly. Hence it should not be left in that condition and should be protected quickly. A final etching or abrasion of the metal surface often helps better adhesion and improved appearance of coating.

Following are the main coatings, generally used for corrosion protection:

(*a*) Metallic coating

(*b*) Inorganic non-metallic coating

(*c*) Organic coating:

(*i*) paints

(*ii*) Varnishes

(*iii*) Enamels

(*iv*) Lacquers

(*v*) Emulsion paint and

(*vi*) Special points

6.6.3.1 Metallic Coatings

The metal used for coating may be put under two categories:

(*i*) Metals which are anodic to the metal to be protected, *i.e.* stand above it in the galvenic series, or

(*ii*) Cathodic metals, *i.e.*, stand below in the galvenic series.

Metals like Zn, Mg, Cd and Al when applied on ion and steel fall under the first category. These being anodic get themselves corroded, saving the iron which is to be protected. As explained under 'formation of anodic areas' in such cases even if the surface is scratched the metal to be protected does not get corroded. These metals may be applied on the surface by hot dipping or spraying.

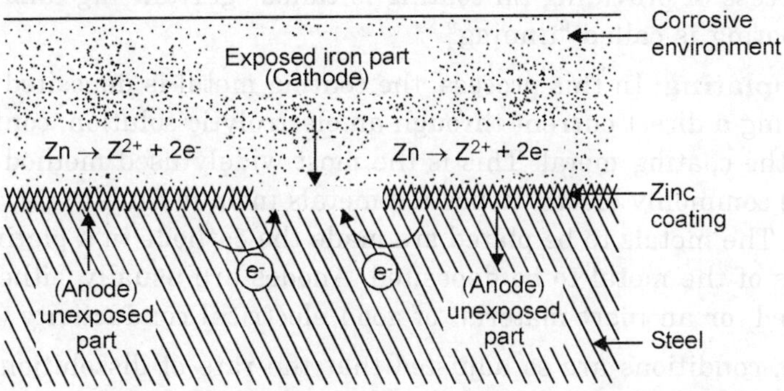

Fig. 6.6. Functioning of anodic coating.

Metal like Sn, Ni and Cu on iron and steel surfaces fall in the second categories. These being cathodic to iron are more resistant to corrosion and offer protection only and so long as the surface is completely covered. If the coating breaks or is applied in a manner that pinholes are left, these points at which the iron surface is exposed get very severely corroded because they are anodic to the large cathodic surface. Hence great care has to be exercised while applying these.

Generally, the cathodic metal coating is applied by electroplating because by this method a more uniform coating which completely cover the surface is produced.

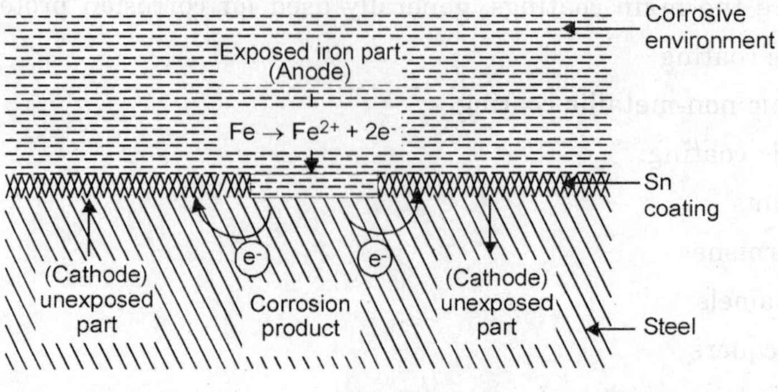

Fig. 6.7. Functioning of cathodic coating.

Metallic coating may be applied in any one of the following ways:

(*i*) Hot dipping

(*ii*) Electroplating

(*iii*) Metal spraying

(*iv*) Metal cladding and

(*v*) Cementation

(*i*) **Hot dipping:** In this process, the metal to be coated is dipped in the molten bath of the coating metal for sufficient time and then removed out along with the adhering film. Metals like Zn, Sn and Pb and alloys like solder and terneplate (Ph-Sn alloy) are used for coating by this method.

The process of providing Zn coating is called "galvanising" and the one providing a Sn coating is called "tinning".

(*ii*) **Electroplating:** In this process, the coating metal is deposited on the base metal by passing a direct current through an electrolytic solution, containing the soluble salt of the coating metal. This is the most widely used method of coating metals; and the commonly electro deposited metals include Au, Ag, Cu, Sn, Ni, Cd, Cr, Zn, Pb, etc. The metals to be plated are made the cathode in a plating bath containing the ions of the metal to be deposited. Anodes are usually, either the metals to be deposited, or an inert material of good electrical conductivity (graphite).

Usually, conditions are so adjusted that the rate of dissolution at the anode and the rate of deposition at the cathode are almost equal, so that the composition of the electrolyte remains unchanged.

(*iii*) **Metal spraying:** Metal spraying is the process by which, the coating metal (like Al, Zn, etc.) in the molten state is sprayed on the surface of the base metal by spraying gun or pistol. The metal thus sprayed adheres to be the base metal surface. The sprayed metallic coatings are somewhat more porous than those obtained by other processes. One of the great advantage of this method is that this can be applied to the finished structures such an bridges, tanks, ships, towers, etc. and so damage to the coating during assembly is avoided.

(iv) **Metal cladding:** In the method, the metal or alloy to be protected is sandwiched between two layers of the protecting metal, and the entire combination is pressed together by rolling, *e.g.* Duralumin, an aluminium alloy containing 4% Cu, 0.5% Mg, 0.5% Mn and small amount of Si and Fe, cladded with aluminium produces a material which combines the strength of the alloy with the corrosion resistance of the pure metal. This type of coating is used in the equipment for handling chemicals, petroleum and in rubber, paper and soap industries, gasoline and water tanks, etc.

(v) **Cementation:** Cementation is also known as diffusion coating. In this process, the base metal is heated with the coating metal in the form of powder to a sufficiently high temperature, in order to promote diffusion of the coating metal into the base metal. This method is used for producing alloy layer on iron and steel surfaces with Zn (sherardizing), Al (colorizing), Cr (chromizing), Si (siliconizing), etc.

Following are the important metallic coatings:

(i) *Zinc coatings* are generally obtained by hot dipping of the base metal in a molten Zn bath and the process is called galvanizing.

(ii) *Tin coatings* are obtained by hot dipping or electrodeposition from acid stannous sulphate or from alkaline sodium stannate baths, or by spraying.

(iii) *Nickel coatings* are usually applied on iron, steel and brass articles mostly by electrodeposition.

(iv) *Chromium coatings* are non-tarnishing, lustrous and decorative and are widely used for automobiles, hardwares, radiator-grillers, metal furnitures, etc.

(v) *Aluminium coatings* on steel are generally applied by spraying, hot dipping, or cementation and are useful in preventing scaling oxidation of ferrous materials even at high temperatures.

(vi) *Copper and lead coatings* are also used for some limited applications.

6.6.3.2 Inorganic Metallic Coatings

Inorganic metallic coatings are also known as non-metallic coatings. The inorganic non-metallic protective coatings include:

(i) Surface conversion or chemical conversion coating or ceramic coatings

(ii) Anodized oxide coatings and

(iii) Vitreous enamel coating.

(i) **Surface or chemical conversion coating:** These coatings are produced by covering the surface of a metal or alloy by chemical or electro-chemical methods. In this process, the metal is immersed in a solution of a suitable chemical which reacts with the metal surface producing an adherent coating. These coatings afford good protection of the base metal from corrosion in some environments and sometimes are of decorative value. Further, many of these coatings are particularly useful to serve as excellent bases for the application of paints, enamels and other organic protective coatings.

The most commonly used surface conversion coatings are chromate coatings, phosphate coatings and oxide coatings.

Chromate coatings are applied on Al, Zn, Mg, etc. by immersing first in acid chromate bath followed by that of neutral chromate.

Oxide coatings can be obtained by treating the metal with alkaline oxidizing agents. Protective films of magnetic oxide on iron can be obtained by passing steam over the heated metal.

Phosphate coatings may be obtained by reacting the base metal with phosphoric acid and phosphates of Zn, Fe or Mn in the presence of a copper salt which acts as an accelerator. The phosphating processes frequently employed are parkerising, bonderizing and cosletizing. These are patented processes.

(ii) **Anodized oxide coating:** Anodized oxide coatings are, generally, produced on non-ferrous metals like Al, Zn, Mg and their alloys by anodic oxidation process, in which the base metal is made as anode. It is carried out by passing a moderate direct electric current through a bath in which the metal or alloy is suspended from anode. The bath usually contains sulphuric, phosphoric, chromic, oxalic or boric acid. After anodizing, the oxide coating is 'sealed' by immersing in boiling water.

As the anodized coatings are somewhat thicker than the natural oxide films, so they possess improved resistance to corrosion as well as mechanical injury.

(iii) **Vitreous enamel coating:** Vitreous or porcelain enamel coatings are modified glass-like materials having different compositions which are usually applied on steel and cast iron equipment. The metal part to be enamelled is first cleaned carefully to remove grease and oxide scale. The vitreous material for the enamel, called "frit" is prepared by fusing together refractory acidic substance with basic fluxes (*e.g.* borax, cryolite, soda ash, feldspar, etc.). The 'frit' is applied to the metal by any of the following methods:

Wet process: In this process, the frit is ground in a pebble mill with water, floating agent (*e.g.* clay), opacifier (*e.g.* Sn, Zn or Sb_2O_3) or colouring agent.

The mixture is called the 'slip' to which electrolyte such as soda-ash or borax may be added before application for proper 'set up' on the metal surface. The metal part which is to be enamelled is dipped in the slip. The slip may also be sprayed over the metal part.

Dry process: In this process the frit is ground in a dry mill together with opacifiers or colouring agents. The finally ground mixture is dusted on the heated surface which has an undercoat of a moist adhesive or a ground coat by the wet process.

High proportion of silica and feldspar and lower proportions of fluxes give acid-resistant enamel.

Ceramic protective coatings made from high refractory oxides such as Cr_2O_3 afford protection of the metal from corrosion even at high temperatures.

These coatings are widely applied for ferrous materials used for equipment in the pharmaceutical chemical, dairy, food and beverage industries. Enamelled steel is used for refrigerators, stoves, tabletops, kitchen utensils, etc. Enamelled cast iron is widely used for bath tubes, sanitarywares, pans, basins etc. used in hospitals.

6.6.3.3 Organic Protective Coating

Organic coatings are inert organic barriers (like paints, varnishes, lacquers and enamels) applied on metallic surfaces and other constructional materials for both corrosion protection as well as decoration. The protective value of such coatings depends on:

(*i*) its chemical inertness to the corrosive environment

(*ii*) its good surface adhesion

(*iii*) its impermeability to water, salts and gases and

(*iv*) its proper application method.

(*i*) **Paints:** Paints are a viscous suspension of finely divided solid. One or more pigments in a fluid medium which on drying, yields an impermeable film having considerable hiding or obliterating power. The medium consists of non-volatile film forming materials like drying oils and/or resins in suitable volatile solvent (thinners). When a properly treated metal surface is painted, the thinner evaporates leaving behind the film and the pigment on the surface. The mixture gradually dries into a solid film. The film formation can be accelerated by the addition of driers in case of drying oil, which are metallic soaps.

The important constituents of a paint are:

(*i*) Pigment

(*ii*) Vehicle or medium (drying oil)

(*iii*) Thinner

(*iv*) Drier

(*v*) Fillers or extenders and

(*vi*) Plasticizers

(*ii*) **Varnishes:** Varnishes are solutions of natural or synthetic resins in a vehicle consisting of a drying oil, a thinner and a drier. A varnish does not contain any pigment. It is used as a protective and/or decorative coating of suitable surfaces and dried by evaporation, oxidation and polymerisation of its constituents, leaving behind a hard, transparent, glossy, lustrous and durable film.

Varnishes are of two types:

(*i*) Oil varnish or oleoresinous varnish

(*ii*) Spirit varnish

The properties of a varnish depend upon the nature of the resin and the oil used. Varnishes are used for protection from corrosion. They are mainly used for wooden furniture to improve their brightness.

(*iii*) **Enamels:** Enamels are the pigmented varnishes or lacquers. They contain pigments, oil or resin vehicle, driers and thinners. Enamels give the combined advantages of a paint and varnish and produce harder, smoother, more lustrous and more glossy finish than ordinary paints.

(*iv*) **Lacquers:** Lacquers are solutions of film forming materials like cellulose derivatives, resin and plasticizer in a suitable solvent, *e.g.*, spirit varnishes, solution of shellac in methanol etc. in which the film formation takes place entirely by evaporation of the solvent.

Lacquers are used for protection from corrosion and for interior decoration. They give tough, durable and glossy coating. They are widely used in automobiles.

(*v*) **Emulsion Paints:** These are the dispersions of synthetic resin latex in water containing oil, resin, pigment and other constituents. The latex (rubber) used are styrene butadiene copolymers, PVC, polystyrene, etc.

6.7 DYES

Aye can generally be described as a coloured substance that has an affinity to the substrate to which it is being applied. The dye is generally applied in an aqueous solutio, and may require a mordant to improve the fastness of the dye on the fibre.

Both dyes and pigments appear to be coloured because they absorbe some wave-length of high preferentially. In contrast with a dye, a pigment generally is insoluble, and has no affinity for the substrate. Some dye can precipitated with an inert salt to produce a lake pigment.

The dyes were obtained form animal, negotiable or mineral origine, with no or very little processing. By far the greatest source of dyes has been from the plant kingdom, notably roots, berries, leaves, bark and wood, but only a few have ever been used on a commercial scale.

6.7.1 Organic Dyes

The first human-made (synthetic) organic dye, mauveine, was discovered by William Henry Perkin in 1856. Many thousands of synthetic dyes have since been prepared.

Synthetic dyes quickly replaced the traditional natural dyes. They cost less, they offered a vast range of new colors, and they imparted better properties upon the dyed materials.[1] Dyes are now classified according to how they are used in the dyeing process.

Acid dyes are water-soluble anionic dyes that are applied to fibers such as silk, wool, nylon and modified acrylic fibers using neutral to acid dyebaths. Attachment to the fiber is attributed, at least partly, to salt formation between anionic groups in the dyes and cationic groups in the fiber. Acid dyes are not substantive to cellulosic fibers.

Basic dyes are water-soluble cationic dyes that are mainly applied to acrylic fibers, but find some use for wool and silk Usually acetic acid is added to the dyebath to help the uptake of the dye onto the fiber. Basic dyes are also used in the coloration of paper.

Direct or **substantive dyeing** is normally carried out in a neutral or slightly alkaline dyebath, at or near toiling point, with the addition of either sodium chloride (NaCl) or sodium sulfate (Na_2SO_4). Direct dyes are used on cotton, paper, leather, wool, silk and nylon. They are also used as pH indicators and as biological stains.

Mordant dyes require a mordant, which improves the fastness of the dye against water, light and perspiration. The choice of mordant is very important as different mordants can change the final color significantly. Most natural dyes are mordant dyes and there is therefore a large literature base describing dyeing techniques. The most important mordant dyes are the synthetic mordant dyes, or chrome dyes, used for wool; these comprise some

30% of dyes used for wool, and are especially useful for black and navy shades. The mordant, potassium dichromate, is applied as an after-treatment. It is important to note that many mordants, particularly those in the hard metal category, can be hazardous to health and extreme care must be taken in using them.

Vat dyes are essentially insoluble in water and incapable of dyeing fibres directly. However, reduction in alkaline liquor produces the water soluble alkali metal salt of the dye, which, in this leuco form, has an affinity for the textile fibre. Subsequent oxidation reforms the original insoluble dye.

Reactive dyes utilize a chromophore containing a substituent that is capable of directly reacting with the fibre substrate. The covalent bonds that attach reactive dye to natural fibers make it among the most permanent of dyes. "Cold" reactive dyes, such as Procion MX, Cibacron F, and Drimarene K, are very easy to use because the dye can be applied at room temperature Reactive dyes are by far the best choice for dyeing cotton and other cellulose fibers at home or in the art studio.

Disperse dyes were originally developed for the dyeing of cellulose acetate, and are substantially water insoluble. The dyes are finely ground in the presence of a dispersing agent and then sold as a paste, or spray-dried and sold as a powder. They can also be used to dye nylon, cellulose triacetate, polyester and acrylic fibres. In some cases, a dyeing temperature of 130°C is required, and a pressurised dyebath is used. The very fine particle size gives a large surface area that aids dissolution to allow uptake by the fibre. The dyeing rate can be significantly influenced by the choice of dispersing agent used during the grinding.

Azo dyeing is a technique in which an insoluble azoic dye is produced directly onto or within the fibre. This is achieved by treating a fibre with both diazoic and coupling components. With suitable adjustment of dyebath conditions the two components react to produce the required insoluble azo dye. This technique of dyeing is unique, in that the final color is controlled by the choice of the diazoic and coupling components.

Sulfur dyes are two part "developed" dyes used to dye cotton with dark colors. The initial bath imparts a yellow or pale chartreuse color. This is oxidized in place to produce the dark black we are familiar with in socks and the indigo blue of the common blue jeans.

6.7.2 Food Dyes

One other class which describes the role of dyes, rather than their mode of use, is the food dye. Because food dyes are classed as food additives, they are manufactured to a higher standard than some industrial dyes. Food dyes can be direct, mordant and vat dyes, and their use is strictly controlled by legislation. Many are azoic dyes, although anthraquinone and triphenylmethane compounds are used for colors such as green and blue. Some naturally-occurring dyes are also used.

6.7.3 Other Important Dyes

A number of other classes have also been established, including:

- Oxidation bases, for mainly hair and fur.
- Leather dyes, for leather.
- Fluorescent brighteners, for textile fibres and paper.

- Solvent dyes, for wood staining and producing colored lacquers, solvent inks, coloring oils, waxes.
- Carbene dyes, a recently developed method for coloring multiple substrates

6.7.4 Chemical Classification

By the nature of their chromophore, dyes are divided into:

- Category: Acridine dyes, derivates of acridine
- Category: Anthraquinone dyes, derivates of anthraquinone Arylmethane dyes
- Category: Diarylmethane dyes, based on diphenyl methane
- Category: Triarylmethane dyes, derivates of triphenyl methane
- Category: Azo dyes, based on –N=N- azo structure
- Category: Cyanine dyes, derivates of phthalocyanine
- Category: Diazonium dyes, based on diazonium salts
- Category: Nitro dyes, based on a $-NO_2$ nitro functional group
- Category: Nitroso dyes, based on a –N=O nitroso functional group
- Category: Phthalocyanine dyes, derivates of phthalocyanine
- Category: Quinone-imine dyes, derivates of quinone
- Category: Azin dyes
- Category: Eurhodin dyes
- Category: Safranin dyes, derivates of safranin Indamins
- Category: Indophenol dyes, derivates of indophenol
- Category: Oxazin dyes, derivates of oxazin
- Category: Oxazone dyes, derivates of oxazone
- Category: Thiazin dyes, derivates of thiazin
- Category: Thiazole dyes, derivates of thiazole
- Xanthene dyes, derived from xanthene
- Fluorene dyes, derivates of fluorene
- Category: Pyronin dyes
- Category: Rhodamine dyes, derivates of rhodamine
- Category: Fluorone dyes, based on fluorone

PROBLEMS

1. What is corrosion of metals? Explain the basic reason of metallic corrosion.
2. Describe briefly the consequences of corrosion.
3. Define metallic corrosion. Explain electrochemical theory of corrosion.
4. State and explain pilling—Bedworth rule.

5. Discuss in detail about chemical corrosion.
6. Differentiate chemical and electrochemical corrosion with suitable examples.
7. What is stress corrosion? Give two examples. How can it be controlled?
8. Discuss any two factors which influence the corrosion rate in detail.
9. Why does corrosion occur in steel pipe connected to copper plumbing? Explain.
10. Discuss briefly (*a*) Soil corrosion, (*b*) Galvanic corrosion, (*c*) Pitting corrosion, and (*d*) Crevice corrosion.
11. Explain briefly the various factors influencing corrosion.
12. Silver and copper metals do not undergo much corrosion like iron in moist atmosphere. Explain.
13. Explain methods of prevention of corrosion.
14. Discuss briefly corrosion inhibitors.
15. What is corrosion? Briefly discuss the various methods employed for protection of metals from corrosion.
16. How are metals protected against corrosion by modifying the environment?
17. How will you protect an underground pipeline from corrosion by sacrificial anodic and impressed current cathodic protection methods?
18. What are cathodic and anodic protection for controlling corrosion? Discuss their merits and demerits.
19. Discuss any four methods of corrosion control.
20. Distinguish between anodic and cathodic inhibitors.
21. What are corrosion inhibitors? Explain with examples how anodic and cathodic inhibitors provide protection against corrosion.
22. Explain rusting of iron with the help of electrochemical theory.
23. Discuss briefly passivity and its significance.
24. Discuss briefly various types of corrosion.
25. Distinguish between anodic and cathodic coating.
26. Explain the preliminary treatment given to a metallic surface before coating.
27. Discuss the importance of design and material selection in controlling corrosion.
28. What is dry corrosion? Why do metals get oxidized on exposure to air?
29. Summarize the various laws of oxidation and give examples of metals which obey such laws.
30. Discuss Wagner's theory of oxidation of metals.
31. How does alloying affect the rate of oxidation of a metal?
32. What are the various scales that appear on the surface of iron heated above 575°C? How is this scale formation useful?
33. Explain the terms:

    (*a*) Differential metal corrosion

(*b*) Differential aeration corrosion and

(*c*) Graphitic corrosion.

34. Give an outline of the electrochemical theory of corrosion and explain the mechanism of just formation in

(*a*) acid

(*b*) natural media.

35. Explain the following:

(*a*) Pure zinc granules do not liberate hydrogen, readily from dilute acids, but there is a vigorous evolution of hydrogen if the metal is in contact with Pt or Cu.

(*b*) Amalgamated zinc reacts slowly with dilute acids.

(*c*) Cu is insoluble in dil. H_2SO_4, but dissolves readily if H_2O_2 is added to the mixture.

(*d*) Pure iron does not dissolve readily in conc. HNO_3.

(*e*) If a drop of 3% solution of NaCl containing peroxyl indicator is placed on a piece of iron, the peripheral regions of the drop becomes blue. A ring of rust appears concentrically around the central region.

36. How do the presence of water vapour, acidic vapours and dust particles influence the corrosion?

37. It is safer to use cathodic inhibitors rather than anodic inhibitors. Explain.

38. Explain the use of Evens polarization diagrams in understanding corrosion control.

39. Summarize briefly how with proper choice of materials and suitable design, corrosion can be minimized.

40. Explain stress corrosion with examples. (*VTU–2006*)

41. What are corrosion inhibitors? Explain how corrosion is controlled by using anodic and cathodic inhibitors? (*VTU–2006*)

42. Write a brief note on the effect of following factors on the rate of corrosion:

(*i*) Nature of Metal

(*ii*) Hydrogen over voltage

(*iii*) Relative areas of anode and cathode.

CHAPTER

7

ADHESIVES

7.1 INTRODUCTION

An adhesive can be defined as "any substance which is capable to hold materials together by surface attachment". The bodies on which the adhesive is applied, are known as adherends, while the process of holding the two or more adherends together by adhesive is called bonding and the total system (adherend and adhesive) is called **'bond' or 'joint'.**

A good quality adhesive should, have the following requisites :

(*a*) Degree of tackiness should be high.

(*b*) Should take less time for bonding.

(*c*) High strength of bond setting on drying.

(*d*) Durability should be high.

7.2 CLASSIFICATION OF ADHESIVES

The adhesives are classified based on their chemical nature. The adhesives are broadly classified into several members but only few of them are discussed below:

7.2.1 Natural Resin Adhesives

(*a*) **Shellac resin**: Shellac resin was used to produce scandess lead pipe and it is applied in hot oxidition. It possesses good bond strength, good hardness and low thermal conductivity. It is used for making conveyers, belts, cables etc.

(*b*) **Asphalt:** It is a crude adhesive and is used in bonding fibres, paper, metals etc.

7.2.2 Thermosetting Synthetic Adhesives

The thermosetting synthetic resins undergo some chemical change and become an infusible and insoluble mass on heating and possess good resistance towards: (*i*) moisture (*ii*) heat (*iii*) insects (*iv*) fungi etc.

Some important thermosetting synthetic resins are discussed below:

(*a*) **Phenol formaldehyde resins (PF):** These are formed by the poly condensation between phenol and formaldehyde in the presence of a catalyst (acids or bases). With excess of formaldehyde, a cross-linked three dimensional resin will ultimately result. This resin possesses great adhesive properties. Paper, fabric or asbestos cloth are impregnated with the resols and many layers of the impregnated material are pressed together in a hydraullic press and heated at around 150°C. The resin cures under pressure and bonds the various layers firmly, to form a single sheet of a greater thickness. These are used as core binders in foundries, water proof plywoods, laminates, coating composition, electrical circuits etc.

(*b*) **Urea-formaldehyde resin:** Urea can condense with formaldehyde to form urea formaldehyde (UF) resins. This resin in general is colourless and clearer, hence, objects of lighter shades can be made out of them. Similarly, the tensile strength and hardness are also much better, however, their heat and moisture resistance are definitely lower, compared to the phenolics. They are used in paper industry to improve the wet strength, plywood industry, laminates, articles in aircraft etc.

(*c*) **Epoxy resins:** They have good adhesive properties and are basically polyethers. Epoxy resins are excellent structural adhesives. They have remarkable chemical resistance and stick to a number of substances including metals and glasses. They also have ability of getting cured, without the application of external heat and possess good electrical resistance. They are used in bonding fibres, paper, glass, leather etc.

(*d*) **Polyvinyls (polyvinyl acetate and polyvinyl chloride):** When the polyvinyl alcohol (PVAL) is admixed with polyvinyls, 25% of the PVAL repeating units are left unreacted to provide strong adhesion to glass. Pulyvinyl chloride is insoluble in its own monomer, so it undergoes polymerisation. When polymerisation is carried out under moderate pressure and then vented at the stage of high monomer conversion, a porous 'dry blend' resin is produced. Such resins are able to absorb large amount of a plasticiser before becoming sticky and are used for bonding nonporous surfaces, sealing operations in food industry etc., It is resistant to chemicals, atmosphere and light. They are used in industrial floorings, foams, protting materials for electrical insulations, metallic and ceramic articles etc.

(*e*) **Silicone resins:** The starting materials for producing silicone polymers are alkyl chlorosilanes, aryl chlorosilanes or substituted ester of orthosilicic acid. In such compounds, there are groups which are easily hydrolisable to form hydroxyl group. These resins are thermally very stable. They are available in liquid, waxy and rubbery forms and they have strong power of adhesion. They are water repellant and very stable at elevated temperature. They also have good corrosion resistance. These resins find use in surface coatings, where water repellancy and chemical resistance are of prime concern, *i.e.*, laminates, ceramics, plastics etc.

(*f*) **Polyesters:** Polyesters are prepared by a polycondensation reaction between a dicarboxylic acid and a diol. It has good mechanical strength upto 175°C and is virtually unattacked by many chemicals. They are used in the manufacture of magnetic recording tapes, aluminised sheets for certain special applications etc.

7.2.3 Thermoplastic Synthetic Resins

These resins soften in a reversible 'physical process under the influence of heat and sometimes of pressure and can be moulded into different shapes in this condition. They retain their shapes on cooling. They form soluble, fusible and poor water resistant adhesives. Important members of this resin are:

(*a*) **Cellulose and its derivatives:** Cellulose is a versatile resin which is found in nature in the form of cotton, hemp, jute, flax etc., in abundance. Cellulose has an extremely high degree of crystalinity and a very high melting point. It decomposes before it begins to melt. It is generally resistant to dissolution in several solvents. Cellulose nitrate when admixed with alcohol, is known as **collodion.** It is generally used in cloth and footwear industries.

(*b*) **Cellulose acetate:** Cellulose acetate is cellulose plastic. They have high impact strength, good dimensional stability, excellent compatibility with plasticisers and much lower moisture absorption. They are used in manufacture of automobile components, non-flammable safety cinematographic films, fibres etc.

(*c*) **Ethyl cellulose:** Ethyl celluloses are the ether derivatives of cellulose. Ethyl cellulose is flexible and tough even at very low temperature. It has better mechanical and electrical properties. Both ethyl cellulose and methyl cellulose find application as finishing materials (textile industry) and as paint bases.

(*d*) **Acrylics:** When used in solution, they form perfect adhesive films, which are stable under all conditions of use. However, they are somewhat costly.

7.2.4 Starch Adhesives

These adhesives can be prepared simply by heating a starch suspension in water or diluted HCl at 150°C to 175°C for a short time. They possess good adhesive properties like (*i*) they can be applied in any condition (*ii*) they dry quickly (*iii*) they spread very easily (*iv*) they do not give foul smell and are very cheap. For manufacturing envelopes, stamps, note books and other study materials etc.

7.2.5 Inorganic Adhesives

The inorganic adhesives are available in the form of gel. A gel can be prepared by heating soda-ash mixed with litharge, MnO_2 etc., and is suspended in water. But as the temperature increases, the strength of the gel is reduced and there is a similar effect when it is exposed to the moisture. It is fungus resistant and its bond do not crack on high temperatures. It is used for preparing card board containers, paper boxes, packing kits etc.

7.2.6 Vegetable Glues or Protein Glues

They are generally obtained from casein, albumin, corn, soyabeans peanuts etc. These are finely crushed and then treated with suitable solvent (acid or base) to get jelly like adhesive paste. They possess poor bond strength and are generally used as adhesive for paper works.

7.2.7 Animal Glues

These are prepared from the bones of animals, organic biproducts, and some industrial biproducts. They are first purified by lime, degreased and then treated with HCl. Then the product is hydrolysed with water to get liquor glue and mixed with borax, or other preservatives, then concentrated. Now the product can be used as glue in different state (solid or liquid).

They possess good adhesive properties but less resistant to chemicals, atmosphere, moisture and fungi. They are extensively used in manufacturing furniture, radio-cabinets, card boxes etc.

7.3 CHEMICAL FACTORS INFLUENCING ADHESION

The adhesive action is mainly influenced by the following chemical factors:

1. Polar effect: Polar groups of adhesives are strongly absorbed by the organic containing surfaces, while the non-polar adhesives are not absorbed due to poor bond strength. It can be improved by addition of polar-groups to non-polar adhesives. For example, by addition of maleic acid to polyvinyl acetate or chloride, the adhesive strength is increased considerably.

2. pH effect: Slightly acidic and slightly alkaline adhesives show better strength, while strong acids and alkalis are affecting the adhesive bond strength severly.

3. Degree of polymerisation: The semi solid fractions of resin adhesive exhibit more adhesive strength than highly cured and tough fractions. Getting maximum bond strength of an adhesive depends on its degree of polymerisation that should lie in a range between unpolymerised and highly polymerised fractions.

4. Side chain in adhesive molecule: The bond strength of an adhesive is quite dependent on the chemical nature, length and the complexity of the side chain in adhesive molecule. The greater the chain length, the greater is the bond strength of adhesive. For example, higher aldehyde with phenol possess greater adhesive strength than that of lower aldehyde-phenol resins.

7.4 TECHNIQUE OF APPLICATION OF ADHESIVE

The methods of application of adhesive are based on physical characteristics of the adhesive (Like liquid, powder etc.). The bond strength can also be developed by applying the appropriate and correct methods but the final strength will be dependent on the other important physical factors like pressure, temperature and the lime of contact. Time and temperature actually determines whether full-curing and drying of the adhesive film have taken place or not. Without proper curing and drying the bond will be very weak.

PROBLEMS

1. Define adhesive and discuss the advantages and the limitations of adhesive bonding.
2. Write brief note on different types of adhesives.

3. How the adhesives act on the adherend surfaces? Write the conditions for adhesive action.
4. What are the various physical and chemical factors influencing the adhesive strength?
5. Differentiate between adhesive and adherend.
6. How the adhesive strength can be developed?
7. Explain the different mechanisms of adhesion. Give suitable examples.
8. (*a*) Write short note on natural and synthetic adhesives.
   (*b*) What are adhesives and give uses.
9. Write short notes on :
   (*a*) Thermoplastic synthetic resin.
   (*b*) Epoxy resin.
   (*c*) Cellulose acetate.
   (*d*) Thermosetting synthetic resins.
10. What are the conditions after bonding of adhesives?
11. What are the materials from which an epoxy adhesive is prepared?
12. How the adhesive and adherend are prepared? Write the application of adhesive.
13. What are the effects of temperature and pressure on the adhesive films?
14. How is phenol-formaldehyde used as adhesive?
15. Discuss the role of adhesives in electrical, automobile and aircrafts industry.

CHAPTER

8

LUBRICANTS

8.1 INTRODUCTION

Any substance introduced between two moving/sliding surfaces with a view to redue the frictional resistance between them, is known as 'lubricant' and the process of reducing frictional resistance between moving/sliding surfaces by the introduction of lubricants in between them, is called as 'lubrication'.

When one solid surface is moved over another solid surface, there is a frictional resistance to motion. If the two surfaces were perfectly clean and smooth this frictional resistance, which is primarily due to molecular attraction, would be very great. Thus, in practice when, one solid is moved over another there is wear of the surface and a loss of energy which appears as heat. Both these affects are undesirable.

Most of the lubricants commonly used are liquids, generally oils, but they may be solids or even gases. Graphite is a common solid lubricant and used as a lubricant in high speed spindles. Quantitatively petroleum supplies the largest amount of materials for making lubricants, but lubricants may be drawn from widely different sources, *e.g.*, in cream separators milk acts as a lubricant; in certain very low temperature equipment liquid air may be the lubricant.

8.2 CLASSIFICATION OF LUBRICANTS

On the basis of consistency, lubricants may be classified into the following three categories:

(*i*) Liquid lubricants,

(*ii*) Semi-solid lubricants, and

(*iii*) Solid lubricants.

8.2.1 Liquid Lubricants

Liquid lubricants are also known as lubricating oil. Lubricating oils reduce friction and wear between two moving metallic surfaces by providing a continuous fluid film in between them. A good liquid lubricant must possess: (*i*) low pressure, or high boiling point (*ii*) adequate viscosity for particular service conditions (*iii*) low freezing point (*iv*) high oxidation resistance (*v*) heat stability (*iv*) non-corrosive properties and (*vii*) stability, to decomposition at the operating temperatures.

Liquid lubricants also act as:

(*i*) Cooling medium

(*ii*) Sealing agent and

(*iii*) Corrosion protection.

Liquid lubricants are further classified into following categories:

(*i*) Vegetable oils

(*ii*) Animal oils

(*iii*) Mineral oils

(*iv*) Blended oils

(*v*) Synthetic oils.

Vegetable and Animal oils: The oils obtained from vegetable and animal sources consist of glycerides (esters with glycerol) of higher fatty acids. These decomposed on heating and do not distil, hence are called 'fixed oils'. These were the principal sources of lubricants before the advent of petroleum oils. They are still used to a comparatively smaller extent for compounding petroleum oils and for making greases. The fixed oils have higher oiliness which gives them lower friction and a higher load carrying capacity than the petroleum oils. They are, however, more expensive, have a greater tendency to oxidation and decomposition at elevated temperatures. Only oils of non-drying group are used for lubrication.

About 5% of oils like castor oil, sperm oil, lard, etc. or their fatty acids are added to petroleum oils to improve their properties. Such mixtures are known as compounded oils and are used as steam cylinder oils, marine engine oils, etc.

Following are the main vegetable and animal oils:

Vegetable oils: Olive oil, palm oil, castor oil, etc.

Animal oil: Whale oil, lard oil, tallow oil, etc.

Mineral oils: Mineral oils are also known as petroleum oils. Mineral oils are obtained from the fractional distillation of petroleum. The length of hydrocarbon chain in petroleum oils varies between 12–50 carbon atoms. The shorter chain oil have lower viscosity than longer chain hydrocarbon.

Mineral oils, generally, consists of a mixture of paraffinic, naphthanic and aromatic hydrocarbons. The physical properties such as viscosity, depends on the relative proportions of these components. Paraffinic oils generally have low viscosity and density, but tend to get oxidized more easily under the working conditions. The aromatics are relatively stable to oxidation, but form insoluble black sludges in high temperature service.

Mineral oils are the most widely used lubricants, because they are: (*a*) cheap (*b*) available in abundance and (*c*) quite stable under service conditions. However, they possess poor oiliness as compared to that of animal and vegetable oils. The oiliness of the petroleum oil can be increased by the addition of high molecular weight compounds like oleic acid, stearic acid, etc. The main mineral oils are petroleum fractions.

Blended oils: It is also known as compounded oils. Desirable characteristics of lubricating oils can be improved by adding small quantities of various additives known as addition agents or specific agents. The oils thus obtained are known as blended oils or compounded oils.

The requirements of modern machinery are such that petroleum oils by themselves fail to lubricate them satisfactorily. Addition of fixed oils improve them but even this is inadequate. This technique of adding other substances for improving the properties has led to extensive research which is developing ever newer substances called addition agents.

Specific agents or combinations have been developed which improve one or more of the desired properties. Some of the typical, properties that are improved by incorporating addition agents are as follows:

(*i*) Extreme pressure

(*ii*) Viscosity index

(*iii*) Oxidation

(*iv*) Corrosion

(*v*) Pour point

(*vi*) Deposit formation

The outline given above shows how the properties of petroleum oil can be improved. As a matter of fact most of the lubricating oils in the market have one or more of the additives. The term compounded or blended oils have been used for oils in which other substances, mainly fatty materials, are added. Mineral oils with various additives are the examples of blended oils.

Synthetic oils: Synthetic oils are oily liquids which are not found naturally or not produced directly during the normal manufacturing and refining process of the petroleum industry.

In many machines, operating conditions are so sever that petroleum lubricants fail to serve the purpose, *e.g.*, in-die casing, furnace equipment, hot rolling mills, etc., the temperature, and the pressure to which the lubricant is exposed are very high and the petroleum and fatty oils will decompose under these conditions. The oils for the bearings of a high altitude jet plane may be subjected to as very high temperature (250°C). The same oil may have to be pumped at –50°C.

Under both conditions the oil has to maintain almost the same properties. To meet such exacting requirements viscous fluids have been synthesized from various organic and inorganic materials. The synthetic lubricants are quite expensive in some cases may be a 100 times as expensive as petroleum lubricants.

The more widely used synthetic lubricants are silicones, fluolubes, phosphate and esters, etc. The distinguishing features of these may briefly be stated as follows:

Inflammability: The silicones, fully chlorinated compounds and some phosphates are non-inflammable.

Flash point: The esters, though inflammable, have relatively high flash point.

Thermal Stability: The silicones and fluorocarbons do not decompose on exposure to high temperature.

Viscosity: Many of the synthetics have a high viscosity index, *i.e.*, the change in viscosity with temperature is small.

The synthetic lubricant like silicone fluids have a chain of silicon atoms in which silicon atoms are joined to each other through oxygen atoms. It is this structure that makes them stable at high temperatures and resistant to oxidation. Various organic groups are attached to the silicon atoms satisfying the remaining bonds. The nature and size of these groups control physical properties like, viscosity, viscosity index, volatility, flash and fire points, oiliness etc.

Silicones are the products obtained by hydrolysis and polymerization of organochlorosilanes *e.g.*, (CH_3), $SiCl_2$ and $(CH_3)SiCl_3$. The general formula of silicones is

$$-R_3SiO\left[\begin{array}{c} R \\ | \\ -Si-O- \\ | \\ R \end{array}\right]_n SiR_3$$

where R is an organic radical ($-CH_3$ or $-C_6H_5$).

8.2.2 Semi-solid Lubricants

The most widely used semi-solid lubricants are grease and varnishes. Lubricating grease is a semi-solid, consisting of a soap dispersed throughout a liquid lubricating oil. The liquid lubricants may be a petroleum oil or even a synthetic oil and it may contain any of the additives for specific requirements.

Greases are prepared by saponification of fat such as fatty acid with alkali like lime, caustic soda etc., followed by adding hot lubricating oil while under agitation. The total amount of mineral oil added determines the consistency of the finished grease. The structure of lubricating greases is that of a gel. Soaps are gelling agent, which give an interconnected structure (held together by intermolecular forces) containing the added oil. At high temperatures, the soaps dissolved in the oil, where upon the interconnected structures ceases to exist and the grease liquifies. On storage grease have a tendency to separate into oil and soap. A grease is a petroleum oil made semi-solid by mixing with it a sodium, lithium or aluminium soap. Most of the common greases are calcium, made with soap thickness. The soap content is generally between 7 to 18% but in special cases it may be as high as 50% or as low as 3%. In a grease, soap has several functions:

(*i*) It acts as thickening agent.

(*ii*) It sticks to the metal surfaces more firmly.

(*iii*) Its nature decides the temperature upto which the grease can be used.

(*iv*) Its consistency and its resistance to water and to oxidation.

Lubricating greases are employed in the following situation:

(*a*) When a machine is worked at slow speed and high pressure.

(*b*) In situations where spilling or spurting oil from the bearings is detrimental to the product being manufactured as in the case of textile mills, paper and food product manufacture, etc.

(*c*) In situations where oil cannot be maintained in position due to bad seal or intermittent operation.

(*d*) In situations where the bearing has to be sealed against entry of dirt, water and grit.

The greases are named after the soap used in their manufacture, *e.g.* calcium, sodium, lithium, aluminium, barium, etc., base grease.

Calcium–soap grease: It is also known as cup-grease. Calcium greases are emulsions of petroleum oil with calcium-soaps, generally prepared by adding requisite amount of $Ca(OH)_2$ to a hot oil under agitation. They have good resistance to displacement by water and are suitable for lubricating water pumps, tractors, caterpillar trades, etc. Calcium-soap grease cannot be used above 65°C temperature, because they deteriorate losing combined water.

Sodium-soap grease: Sodium-soap grease are petroleum oils, thickened by sodium soap. They are not water resistant because the sodium soap content is soluble in water. However, they can be used upto 175°C. They are suitable for use in ball bearings, where the lubricant gets heated due to friction.

Aluminium-soap grease: They usually contain about 5% more mineral oil than calcium or lime base greases of same consistency. Due to lower soap constant they are relatively water proof and have slightly higher dropping points than the calcium base greases. These greases cannot be used beyond 90°C as they tend to become hard and coarse.

Lithium base greases: These are the petroleum oils, thickened by mixing lithium soaps. They are water resistant and suitable for use at low temperatures (upto 15°C) only. They can be used for aircraft applications.

Barium base greases: These are characterised by extreme resistance to removal from bearings by water, good cohesiveness and adhesiveness, high melting point and maintenance of consistency during the service. They are widely used as a multipurpose lubricants for automobiles and from equipment.

Axle greases (Rosin soap grease): Rosin soap greases are prepared by adding lime or any heavy metal hydroxide to resin and fatty oils. The mixture is thoroughly mixed and allow to stand when grease floats as stiff mass. Fillers like talc and mica are also added to them. They are water resistant and suitable for less delicate equipment working under high loads and at low speeds.

8.2.3 Solid Lubricants

The use of solid lubricants become necessary where it is not possible to maintain a fluid film or the presence of a fluid or semi-fluid is undesirable (as in commutator brushes for electric generators and motors), or the temperature or load is too high even for a semi-solid lubricants. An ideal solid-film lubricant should have the following characteristics:

(*i*) Low shear strength

(*ii*) Strong adhesion to the applied surface

(*iii*) Good thermal conductivity

(*iv*) Chemical inertness and

(*v*) Stability at operating temperatures.

Inorganic solid lubricants: These are mainly laminar solids, *i.e.*, layer lattice in solids in which the bonding between atoms within a layer is strong, being covalent or ionic, while those between adjacent layers are relatively weak-arising from Van der Waals forces, *e.g.*, M_2S_2, WS_2, mica, BN, Ag_2SO_4 $CoCl_2$, $CdCl_2$, PbI_2, borax, graphite, etc.

The two most usual solid lubricants employed are graphite and molybdenum disulphide. Graphite consists of a multitude of flat plates, one atom thick, which are held together by only weak-bonds so that the force to shear the crystals parallel to the layers is low. Consequently the parallel layers slide over one another easily. Usually some organic substances are mixed to solid lubricants so that they may stick firmly to the metal surface.

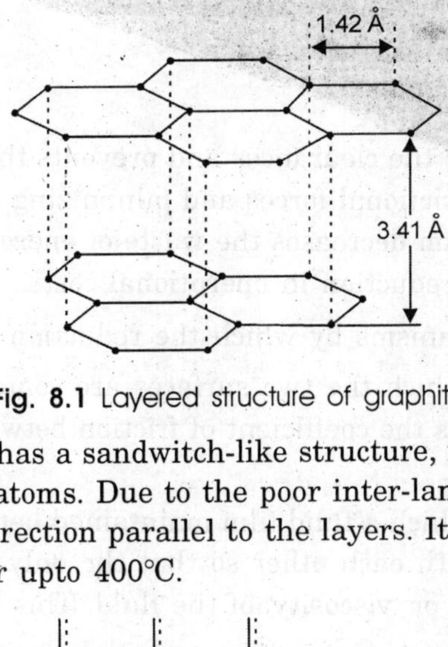

Fig. 8.1 Layered structure of graphite

Molybdenum disulphide has a sandwitch-like structure, in which a layer of a Mo atom lies between two layers of S atoms. Due to the poor inter-laminar attraction is responsible for low sheer strength in a direction parallel to the layers. It possesses very low coefficient of friction and is stable in air upto 400°C.

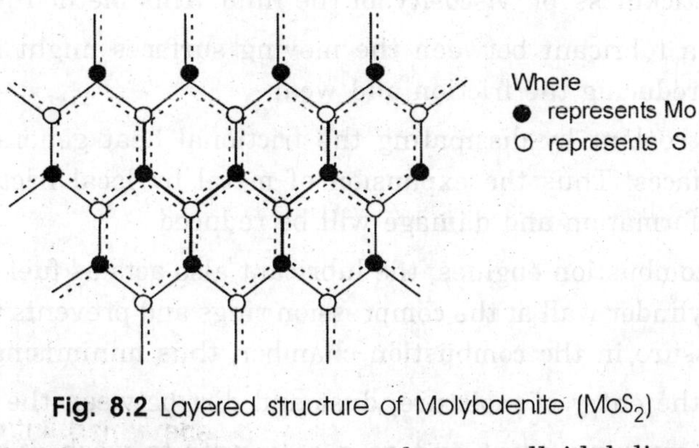

Fig. 8.2 Layered structure of Molybdenite (MoS_2)

Graphite is used either as grease or as powder; as colloidal dispersions (Deflocculated Acheson Graphite, DAG) in water or in oil. When graphite is mixed with water is termed as aqua DAG, whereas when mixed with oil, termed as oil DAG. Graphite is stable in air upto 375°C but above this temperature it oxidises. In the absence of air it can be used upto very much higher temperature. The usual coefficient of friction between solid lubricants is between 0.005 and 0.01.

Graphite is used as lubricants in air compressors, lathes general machine shop works, foodstuff industry, railway track joints, open gears, chains, cast iron bearings, internal combustion engines, etc.

8.3 THEORIES OF LUBRICATION

Lubrication may be defined as the reduction of friction and wear between two relatively moving surfaces by the inter-position of some other substance between the surfaces. The substance thus introduced is known as lubricant.

When a surface slides over another there are three basic physical factors which can affect the overall wear between them:

(*i*) the distance between the surfaces

(*ii*) the force acting on the surfaces

(*iii*) surface texture.

The lubricant is forced into the clearances and prevents the surfaces from meeting and engaging thus decreasing the frictional forces and minimizing the wear and tear. Thus, the presence of a lubricating medium decreases the waste of energy, surface wear, damage and deformation and brings about reduction in operational costs.

There are two basic mechanisms by which the reduction of friction may be achieved.

(*a*) Solid lubrication in which the two surfaces are coated with a substance such as graphite, which lowers the coefficient of friction between the two surfaces as they slide over each other.

(*b*) Fluid lubrication in which a fluid film maintained between the two surfaces keeps them from contact with each other so that the only resistance to motion is that due to the 'stickiness' or 'viscosity' of the fluid. This method is most widely used.

The presence of a lubricant between the moving surfaces might serve several other functions apart from reducing the friction and wear.

(*i*) It acts as a coolant by dissipating the frictional heat generated because of the rubbing surfaces. Thus the expansion of metal by local frictional heat and the resultant deformation and damage will be reduced.

(*ii*) In internal combustion engines, the lubricant also acts as fuel gasket between the piston and cylinder wall at the compression rings and prevents the leakage of gases at high pressure in the combustion chamber, thus minimising the power loss.

(*iii*) It prevents the entry of moisture, dust and dirt between the moving parts.

(*iv*) It acts as a cleaning agent and as a scavenger to wash of and transport solid particles produced in combustion or wear.

(*v*) In aircraft the lubricating oil may be used as a hydraulic fluid to change the pitch of the propeller or to operate other mechanisms. The viscosity index of the oil is the important property in such cases.

When one solid surface is moved over another, there is frictional resistance to motion. Frictional resistance arises due to molecular attraction. When two surfaces were perfectly smooth and clean, this frictional resistance is very high. Even the plane surfaces with which one deals in actual practice are not perfectly plane and smooth. These two have minute projections which interlock each other, and also the surfaces are not perfectly clean.

There may be even on the so-called 'clean' surface, some adsorbed moisture, dirt, etc. such impurities tend to reduce the friction. Thus, in practice when one solid moves over another there is wear of the surface and a loss of energy in the form of heat. This can be avoided by introducing a lubricant (solid, liquid or semi-solid) between the surfaces.

8.4 MECHANISM OF LUBRICATION

Lubrication mechanism can be categorised into following three types:

(*i*) Fluid film or hydrodynamic lubrication

(*ii*) Thin film or boundary lubrication

(*iii*) Extreme pressure lubrication

8.4.1 Fluid Film or Hydrodynamic Lubrication

In such a case, the moving surfaces are separated from each other by a fluid film of lubricants (atleast 1000 Å thick), so that the surface to surface contact and welding rarely occurs. Under hydrodynamic conditions of lubrication, the coefficient of friction will be as low as 0.0001 to 0.03. Hydrodynamic friction occurs in the case of a shaft running at a fair speed in a well lubricated bearing, with not too high a load, a thin film of the fluid lubricant covers the shaft as well on bearing surfaces as indicated in the Fig. 8.3, and the metal surfaces do not come into contact with each other, thus the resistance to movement is only due to the internal resistance of the lubricant. Delicate instruments, light machines like watches, clocks, guns, sewing machine, scientific instruments, etc. are provided with this type of lubrication.

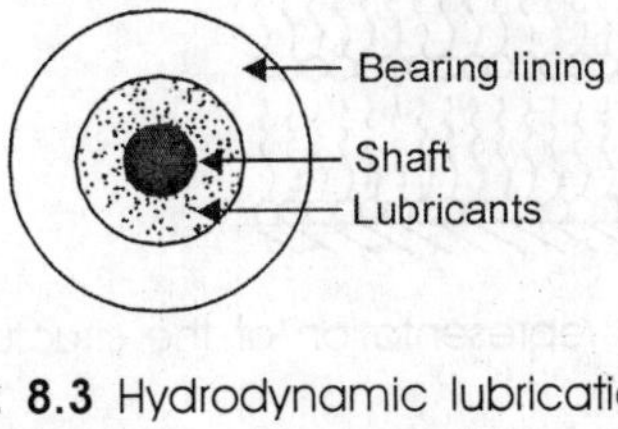

Fig. 8.3 Hydrodynamic lubrications

Hydrocarbon oils are considered to be satisfactory lubricants. Their viscosity increases with the increasing molecular weight. In order to maintain the viscosity of oil, ordinary hydrocarbon lubricants are blended with selected long chain polymers. Hydrocarbon petroleum fractions generally contain small quantities of unsaturated hydrocarbons which will oxidize under operating conditions, forming gums or lacquers. Thus, anti-oxidants, such as aminophenols must therefore be blended with these oils.

8.4.2 Thin Film or Boundary Lubrication

When two solid surfaces are effectively separated by the oil film, and the real contact and sliding takes place along the tails of the hydrocarbon portion sticking out of the solid surface. The longer hydrocarbon chain, greater is the separation between the surfaces and a consequent decreased coefficient of friction. Under heavy loads, the layer of the lubricants may be two or three molecules thick, and under these conditions 'boundary lubrication' occurs. Some of the asperities on the surfaces may be higher than the film thickness and deformation of the asperities occur, leading to some adhesions. The lubricant is displaced at these points of contact. These welds are responsible for the wear observed in this type of lubrication.

Boundary lubrication involves a shearing of both the lubricant film and also the tiny, metallic junctions. The boundary lubrication does not depend on the viscosity of the oil but depends on the load, the oiliness of the lubricants and the properties of the surfaces concerned.

For boundary lubrication, the lubricant molecules should have:

(i) Long hydrocarbon chain.

(ii) Polar groups to promote spreading and orientation over the metallic surface at high pressure.

(iii) Lateral attraction between the chain.

(iv) Active groups or atoms, which can form chemical linkages with the metals or other surfaces.

Boundary lubrication prevails under conditions of high loads, low viscosity, slow moving surfaces and also at the starting and stopping of machinery.

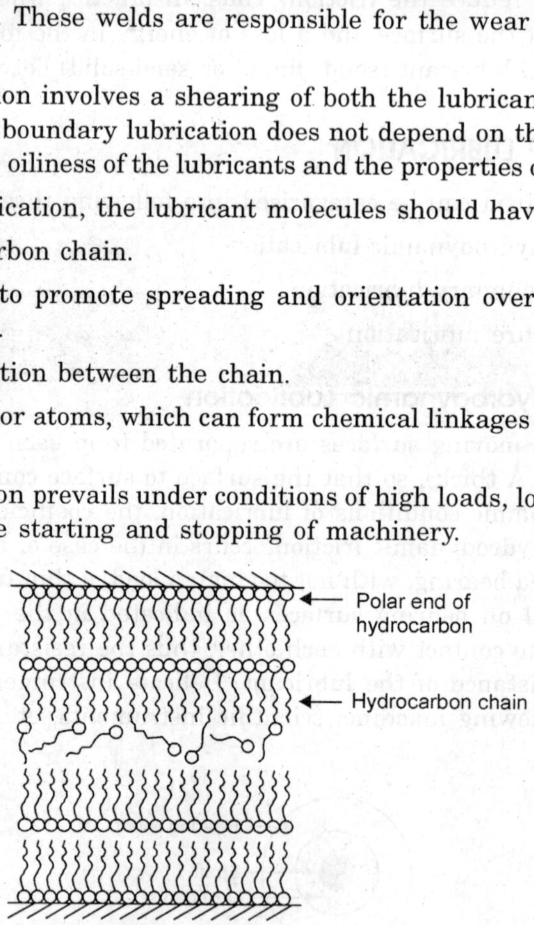

Fig. 8.4 Diagrammatic representation of the structure of adsorbed oil film

8.4.3 Extreme Pressure Lubrication

When the moving solid surfaces are under very high pressure and speed, the lubricants may decompose or vaporise or they may not stick to the surfaces because high local temperature is produced. To meet these extreme pressure conditions, special additives called 'extreme pressure additives' are added to the mineral oil. Extreme pressure additives are the compounds having active groups such as chlorine (chlorinated ester), sulphur (sulphurised oil) and phosphorus (tricresyl phosphate). At high temperature, they react with metal giving surface layers like metallic chlorides, sulphides or phosphides. These surface layers have high melting point (650-1100°C) and serve as good lubricants under extreme pressure and temperature conditions.

SOLVED NUMERICALS

1. An oil of unknown viscosity index has a saybolt universal viscosity of 60 seconds at 210°F and of 600 seconds at 100°F. The high viscosity index standard (*i.e.*, pennsylvanian) oil has saybolt viscosity of 60 seconds at 210°F and 500 seconds at 100°F. The low viscosity index standard (*i.e.*, Gulf) oil has a saybolt universal viscosity of 60 seconds at 210°F and 800 seconds at 100°F. Calculate the viscosity index of unknown oil.

Solution.

$$\text{Viscosity index (VI) of unknown oil} = \frac{V_L - V_X}{V_L - V_H} \times 100$$

where,

$$V_L = 800 \text{ s}$$
$$V_H = 500 \text{ s}$$
$$V_X = 600 \text{ s.}$$

Thus,

$$VI = \frac{800 - 600}{800 - 500} \times 100 = \frac{200}{300} \times 100$$
$$VI = 66.67.$$

2. An oil sample under test has a saybolt universal viscosity of 64 seconds at 210° F and 564 seconds at 100° F. The low viscosity standard (Gulf oil) possesses a saybolt viscosity of 64 seconds at 210° F and 774 seconds at 100° F and 414 seconds at 100° F. Calculate the viscosity index of the oil sample under test.

Solution.

$$VI = \frac{V_L - V_X}{V_L - V_H} \times 100$$
$$= \frac{774 - 564}{774 - 414} \times 100$$
$$= 58.33$$

3. How much rust ($Fe_2O_3.3H_2O$) will be formed, when 100 kg of iron have completely rusted away?

Solution.

214 gm of rust (mol. wt.) contains = 112 gm of iron

∴ 112 gm of iron products rust = 214 g

$$\text{or} \quad 100 \text{ kg of iron produces rust} = \frac{214 \times 100}{100} \text{ kg}$$

$$= 191.1 \text{ kg.}$$

4. Calculate the acid value of a lubricating oil, whose 100 gm required 5 ml of N/20 KOH for complete neutralization.

Solution.

$$\text{Acid value} = \frac{\text{No. of ml of N/10 KOH used} \times 5.6}{\text{Mass of Oil in gm}}$$

$$5 \text{ ml of N/20 KOH} = 5 \times \frac{10}{20} \text{ ml of N/10 KOH}$$

$$= 2.5 \text{ ml of N/10 KOH}$$

Thus,

$$\text{Acid value} = \frac{2.5 \times 5.6}{100}$$

$$= 0.14$$

5. 10 gm of an oil is saponified with 25 ml of N/2 alcoholic KOH. The mixture then required 15 ml of N/2 HCl. Calculate the saponification value of the oil.

Solution.

$$\text{Saponification value} = \frac{\left[\begin{array}{c}\text{No. of ml of} - \text{No. of ml of N/2 acid} \\ \text{N/2 KOH taken used for back titration}\end{array}\right]}{\text{Mass of oil (in gm) taken for saponification}} \times 28$$

$$= \frac{[25 - 15]}{10} \times 28$$

$$= 28.0.$$

PROBLEMS

1. Define a lubricant. Discuss the classification of lubricants with suitable examples.
2. What are solid lubricants? When can they be used? Give details about graphite.
3. Discuss the important properties of lubricants and indicate the significance of these properties.
4. What are the different synthetic lubricants used? How are they superior to petroleum lubricants?
5. Discuss the theory of lubrication.
6. Explain the following:
   (*a*) Lubricant
   (*b*) Asperities
   (*c*) Tribology and
   (*d*) Coefficient of friction.

7. What are the various mechanisms of Lubrication?
8. What are the types of lubricants that are likely to be useful when these mechanisms are in operation?
9. How are the lubricants classified? Give examples.
10. What are the important properties that an efficient lubricant should possess?
11. 1.5 gm of a lubricant oil dissolved in alcohol require 24 ml of 0.0125 N alcoholic KOH for neutralization. What is the neutralization value (NV) of the lubricant? If an oil to be used in machinery should not have NV greater than 8.0, is it safe to use the test sample as a lubricant in the machinery?
12. Why should the following be added to lubricants?
    (*a*) antioxidants
    (*b*) corrosion inhibitors
    (*c*) pour point depressants and
    (*d*) VI improvers. How are these believed to function?
13. Define 'grease' under what conditions should they be used as lubricants?
14. What are the problems associated with the use of gases as lubricants?
15. What are the various factors, that govern the choice of a lubricant?
16. What are the characteristics of a good solid lubricant?
17. How are lubricants in a pure form obtained from the higher boiling fractions of petroleum?
18. Discuss the syntheses and use of the following:
    (*a*) dibasic acid esters
    (*b*) silicons and
    (*c*) polyglycol ethers
19. Explain how the following act as lubricants:
    (*a*) graphite and
    (*b*) molybdenum disulphide
20. What are the various factors that govern the choice of a lubricant?

CHAPTER

9

COLLOIDAL CHEMISTRY

9.1 INTRODUCTION

Thomas Graham studies the ability of dissolved substances to diffuse into water across a permeable membrane. He observed that crystalline substances such as sugar, urea and NaCl passed through the membrane, while others like glue, gelatin and gum did not. The former he called crystalloids and the latter colloids (Greek Kolla = glue; eidos = like).

In a true solution as sugar or salt in water the solute particles are dispersed in a solvent as single molecules or ions. Thus, the diameter of dispersed particles ranges from 1Å to 10Å. On the other hand, in a suspension as sand stirred into water, the dispersed particles are aggregates of millions of molecules. The diameter of these particles is of the order of 2000Å or more. The colloidal solution or colloidal dispersions are intermediate between true solution and suspensions. Thus, when the diameter of the particles of a substance dispersed in a solvent ranges from 10Å–2000Å. The system is termed as colloidal solution or colloidal dispersion or simply a colloid.

1Å   10 Å   2000 Å   More than 2000 Å

True solution	Colloidal solution	Suspension

In colloidal solutions, the particles consists of aggregates of atoms or small molecules *e.g.*, colloidal solution of gold contains particles of various sizes having a million or more atoms. Haemoglobin, a protein molecule responsible for the red colour of the blood, is made up of about 10^3 atoms.

Colloidal systems are heterogeneous in nature and consists of two phases. A substance distributed in a solvent in the colloidal state is said to constitute the dispersed phase and the solvent itself is called the dispersion medium. *e.g.* In the case of a suspension of clay in water, the dispersed phase consists of the clay particles while the dispersion medium is water.

Following are the different types of colloidal systems.

Dispersed Phase	*Dispersion Medium*	*Colloidal Systems*	*Examples*
Solid	Gas	Aerosol	Smokes
Solid	Liquid	Sol./Suspension	Sulphur
Solid	Solid	—	Artificial gems pigmented plastics, alloys
Liquid	Gas	Aerosol	Fog, mist, Cloud
Liquid	Liquid	Emulsion	Oil in water or water in oil
Liquid	Solid	Gels	Jellies
Gas	Liquid	Foam	Whipped cream
Gas	Solid	—	Pumice stone

Colloidal solutions are often referred to as sols. If water is the dispersion medium, they are called hydrosols. Similarly if alcohol and benzene are the dispersion medium, they are called alcosol and benzosols respectively.

9.2 CLASSIFICATION OF COLLOIDAL SYSTEMS

Colloidal systems are usually classified into following two categories:

- Lyophobic (solvent hating)
- Lyophilic (solvent loving)

9.2.1 Lyophobic Colloids

In lyophobic sols there is no affinity between the particles and the solvent. Lyophobic sols. are easily precipitated by the addition of electrolytes. The precipitated sols can not be brought back in the colloidal state and are therefore also called irreversible sols. Most of the inorganic substances form lyophobic sols.

9.2.2 Lyophilic Colloids

Lyophilic sols are more stable than the lyophobic sols. In this case precipitated sols can easily be brought back into the original colloidal state and thus they are called reversible sols. Most of the organic substances form reversible sols. If water is used as solvent, the sols are named as hydrophobic and hydrophilic respectively. The essential difference between hydrophobic and hydrophilic sols are as given:

Properties	*Lyophobic Sols*	*Lyophilic Sols*
Surface tension	Comparable to the dispersion medium.	Usually lower than dispersion medium.
Viscosity	Comparable to the dispersion medium.	Much higher than the dispersion medium
Precipitation	A small amount of an electrolyte causes precipitation.	A small amount of an electrolyte has usually no effect.
Visibility	Particles can be detected by ultramicroscope.	Particle cannot be easily detected by ultramicroscope.
Effect of Electric field	Colloidal particles migrate in one direction depending on the type of charge that they carry.	Colloidal particles may migrate in either direction or may not migrate at all.

9.3 PREPARATION OF COLLOIDAL SOLUTIONS

9.3.1 Preparation of Lyophobic Colloids

The lyophobic colloids can be prepared using following methods.

(*a*) Condensation method, and

(*b*) Dispersion methods.

(a) Condensation Method

In this method, the group of ions or molecules are converted into larger aggregates of insoluble particles of colloidal size. This is generally achieved by means of chemical reactions. Some of the chemical reactions which have been used are as follows:

1. **Reduction:** Sols of metals are usually obtained by reduction of their salt solutions in water by using certain reducing agents. *e.g.,* gold sols is prepared by reduction of auric chloride ($AuCl_3$).

$$\underset{\text{1\% soln.}}{2\,AuCl_3} + \underset{\substack{\text{formalin}\\ \text{2/3 ml}}}{3HCHO} + 3H_2O \xrightarrow{\Delta} 2Au + 6HCl + 3HCOOH$$

2. **Oxidation:** Sols of non-metals are usually obtained by oxidation. When O_2 is bubbled through an aqueous solution of H_2S, it is oxidised and colloidal sulphur is obtained.

$$2H_2S + O_2 \longrightarrow 2H_2O + 2S$$

3. **Hydrolysis:** This method is commonly used for the preparation of hydroxide and hydrous oxide sols. *e.g.,* for the preparation of ferric hydroxide sols, ferric chloride sols is poured into about 750 ml of boiling distilled water. This hydrolysis of ferric chloride occurs instantaneously and a beautiful deep red sol of ferric hydroxide is formed.

$$FeCl_3 + 3H_2O \longrightarrow Fe(OH)_3 + 3HCl$$

4. **Double Decomposition:** This method is generally used for the preparation of sols from insoluble salts. *e.g.,* for the preparation of arsenious sulphide sol, 1–2 gm of pure arsenious oxide is dissolved in 500 ml of boiling distilled water. The resulting saturated solution is diluted with distilled water and to this solution pass H_2S gas carefully. An orange colored colloidal sols of arsenious sulphide is obtained.

$$As_2O_3 + 3H_2S \longrightarrow As_2S_3 + 3H_2O$$

5. **Exchange of solvent:** There are a number of substances whose colloidal solutions can be prepared by taking a solution of the substance in a suitable solvent and then pouring it into another solvent in which the substance is insoluble. *e.g.,* colloidal solution of phosphorous can be prepared by pouring its alcoholic solution in water.

(b) Dispersion Method

In this method, bigger particles are disintegrated into particles of colloidal dimensions by suitable methods. Some dispersion methods are as follows:

1. **Mechanical dispersion:** The bigger particles like graphite are subdivided into fine powder of colloidal size by using mortar. Colloidal mill is used for the preparation of colloidal solution of plastic and rubber.

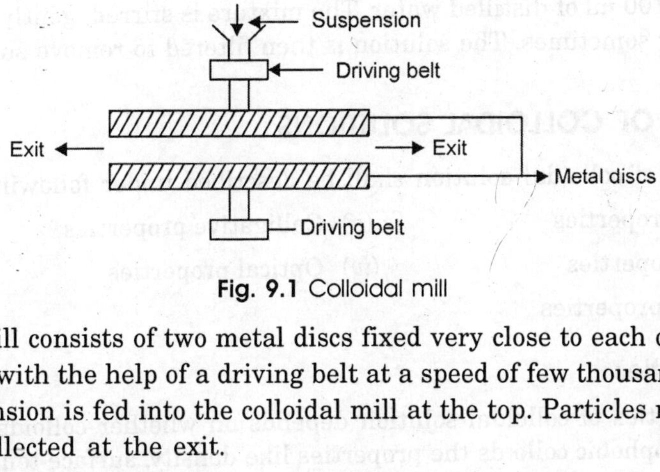

Fig. 9.1 Colloidal mill

A colloidal mill consists of two metal discs fixed very close to each other and moves in opposite direction with the help of a driving belt at a speed of few thousands per revolutions per second. Suspension is fed into the colloidal mill at the top. Particles reduced to colloidal dimensions are collected at the exit.

2. **Electro-dispersion (Bredig's arc method):** This method is mostly used for the preparation of sols of metals such as gold and silver in water. An arc is struck between two rods made up of the material to be dispersed.

A voltage of 300–400 volts is normally used. The electrodes arc immersed in water which contain trace amount of caustic soda. The vessel in which the dispersion occurs is cooled in ice. When an arc is struck between the metal electrode, the metal is vapourised. Condensation of these vapour occurs, as a result of which particles of colloidal size get dispersed in water and a sol of the metal is obtained.

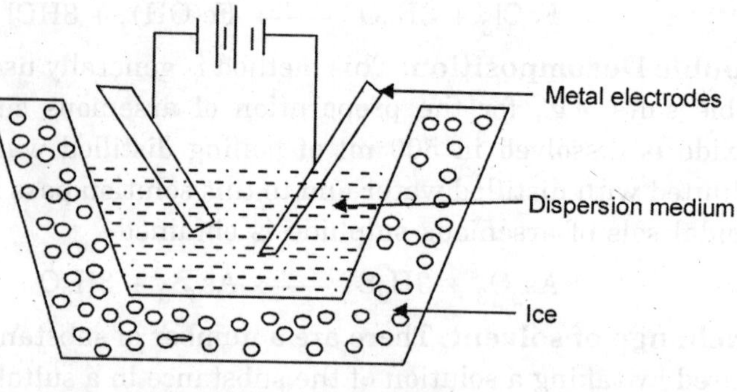

Fig. 9.2 Bridig's arc method

3. **Peptization:** In this method, partial or incomplete dissolution of coarse precipitates results in particles of colloidal dimensions. *e.g.,* when freshly precipitated ferric hydroxide is treated with a small amount of ferric chloride solution, a dark redish brown colored colloidal solution of ferric hydroxide is formed. Here $FeCl_3$ sets as peptizing agent.

9.3.2 Preparation of Lyophilic Colloids

Some substances of complex composition and large molecules such as gelatin, gum arabic, starch, egg albumin, glycogen etc., form colloidal solutions if merely left in contact with water for a sufficient time. To prepare their colloidal solution, 1–2 gm of the substances are added to about 100 ml of distilled water. The mixture is stirred, gently warmed and then allowed to stand for sometimes. The solution is then filtered to remove suspended particles.

9.4 PROPERTIES OF COLLOIDAL SOLUTIONS

The properties of colloidal solution shall be described under following headings:

(*i*) Physical properties
(*ii*) Colligative properties
(*iii*) Kinetic properties
(*iv*) Optical properties
(*v*) Electrical properties

(*i*) Physical Properties

Physical properties of colloidal solution depends on whether colloids are lyophobic or lyophilic. For dil. lyophobic colloids the properties like density, surface tension and viscosity are not very much different from the dispersion medium. The lyophilic colloids show a high degree of solvation and thus the viscosity is much higher than the medium.

(*ii*) Colligative Properties

Colligative properties such as osmotic pressure, elevation of boiling point, depression of freezing point depends upon the M.Wt., of particles present. In colloidal system, the dispersed particles have a very high average M.Wt., often of the order of 10^6. The magnitude of this colligative property is, therefore, usually very small. However, some colloids do exhibit a measurable osmotic pressure. In such cases, osmotic pressure measurement have been used to determine the M.Wt. of colloidal solutions.

(iii) Kinetic Properties

Rober Brown observed a strange and interesting motion of small particles of pollen grains, minute particles of coal, dust and mineral suspended in water. Such a movement was named as Brownian movement. He noted that these particles move continuously and quite irregularly in all directions.

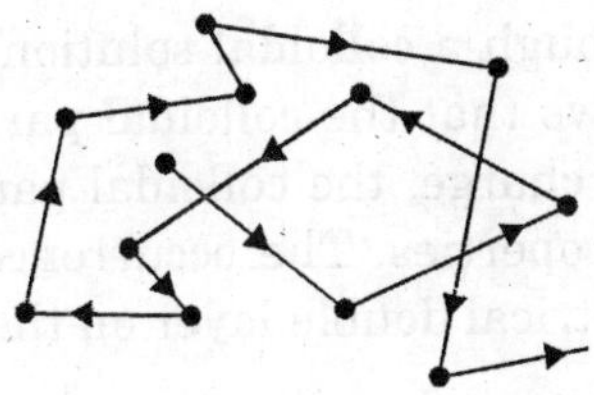

Fig. 9.3

He pointed out that Brownian movement is due to irregular bombardment of the suspended particles by the molecules of the liquid which are in continuous random motion. Quantitative measurements showed that this movement is observed when the particle have diameter not larger than 5μ (0.0005 cm). The smaller the particles size, the more rapid is the movement. The movement depends upon temperature, the density of the movement increases with increasing temperature.

Diffusion: Since the colloidal particles are in chaotic (disorderly) motion, they diffuse and spread out into the solvent. Diffusion is thus, the result of Brownian movement. The rate of diffusion is characterised by diffusion coefficient or diffusion constant. *i.e.*, the weight of material passing in one second across a plane of one cm^2 area when the concentration gradient is unity. The diffusion coefficient is related to the mean free path A of the particles by the following equation:

$$A^2 = 2D.t$$

where, D = diffusion coefficient

t, is the time in which the particles moves the distance A.

According to Brownian movement, "diffusion coefficient is directly proportional to the temperature and inversely proportional to the radius of the particles and to the viscosity of the medium."

$$D = \frac{RT}{N} \cdot \frac{1}{6\pi\eta r}$$

where, T denote temperature, r = radius of the particles, η = viscosity.

(iv) Optical Properties

When a beam of light is passed through a true solution, the path of beam through the solution is invisible. However, when light is passed through a colloidal solution, the beam becomes visible. This is because of the scattering of light by colloidal particles. This scattering was observed for the first time by **Tyndall** and is known as **Tyndall effect.**

The scattering of light occurs due to difference in the refractive indices of the dispersed phase and the dispersion medium. This difference is very well pronounced in the case of lyophobic solutions and hence the Tyndall effect is quite appreciable. In lyophilic solutions, the differences in refractive indices is almost negligible and hence the Tyndall effect is extremely weak. From the knowledge of the volume of the solution and the number of particles contained thereium, the size of particle can be estimated.

(v) Electrical Properties

When a current is passed through a colloidal solution, the particles move either to the +ve or the –ve electrode. This shows that the colloidal particles are electrically charged. As a result of the presence of electric charge, the colloidal particles exhibit a set of interesting properties called electro kinetic properties. The occurrence of these electrical properties is attributed to the existence of a electrical double layer on the surface of the colloidal particles.

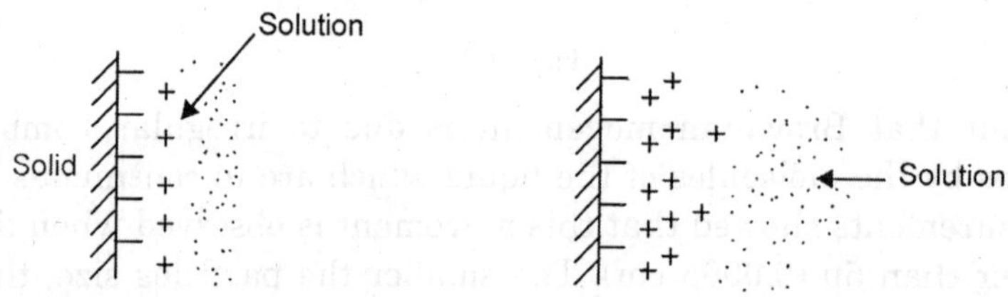

Fig. 9.4 (*a*) Helmholtz double layer **Fig. 9.4** (*b*) Diffused double layer

When a solid comes in contact with liquid, it preferentially adsorbs certain ions. *e.g.*, when water is brought in contact with a glass surface, the glass adsorbs OH^- ions and become negatively charged with respect to water. To balance this charge, H^+ ions are attracted to the surface to form a double layer of charges, with the –ve charge on glass and +ve charges in water immediately adjacent to these. This type of arrangement of charge is called Helmholtz double layer. Thus a difference of electric potential is developed at the solid-liquid interface.

In colloidal solution, a diffuse electrical double layer appears at the surface of the colloidal particles as shown in the figure. The potential across the diffused part of the double layer is called zeta (ξ) potential. This potential is responsible for the occurrence of the so-called electro kinetic phenomena.

Following two types of electro-kinetic phenomena on discussed here:

(*i*) Electrophoresis

(*ii*) Electro osmosis

Electrophoresis

The migration of electrically charged colloidal particles under the influence of an electric potential is called electrophoresis. This phenomenon can be studied by using an experimental arrangement as shown in Fig. 9.5.

A U-tube contains the sol to be studies. A small quantity of dispersion medium (water) is added to either of the two limbs. Distinct boundaries are formed between the colloidal solution and the added dispersion medium. With the help of two electrodes, a potential difference is now applied. It is seen that the boundary starts moving towards one of the

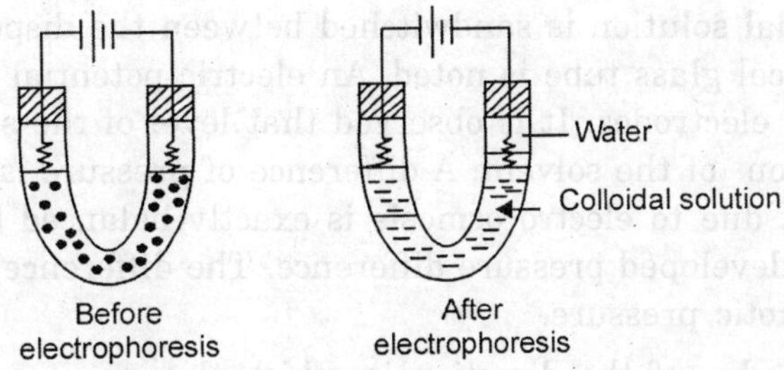

Fig. 9.5

electrodes. The movement of boundary occurs due to the movement of the colloidal particles under the influence of the applied potential difference. If the boundary moves towards the cathode, it may be concluded that the sol is +vely charged and if it moves towards the anode, it is –vely charged. The distance moved by the particle in one second and under a potential gradient of one volt/cm is called electrophoretic mobility (U). Electrophoretic mobility is related to zeta potentials as follows:

$$U = \frac{D}{4\pi\eta}\xi$$

where, D = dielectric constant

η = viscosity

ξ = zeta potential

Some of the +vely and –vely charged solutions are:

+ve: $Fe(OH)_3$, $Al(OH)_3$, Chromium hydroxide, Thorium hydroxide, Zirconium hydroxide.

–ve: Gold, silver, platinum, sulphur, arsenous sulphide V_2O_5 led sulphide etc.

Electro osmosis

If the colloidal particles are prevented from moving, the solvent moves under the influence of the electric potential. This migration of the solvent under the influences of the electric potential is called electro osmosis. Electro osmosis can be demonstrated with the help of an experimental set up as shown in Fig. 9.6.

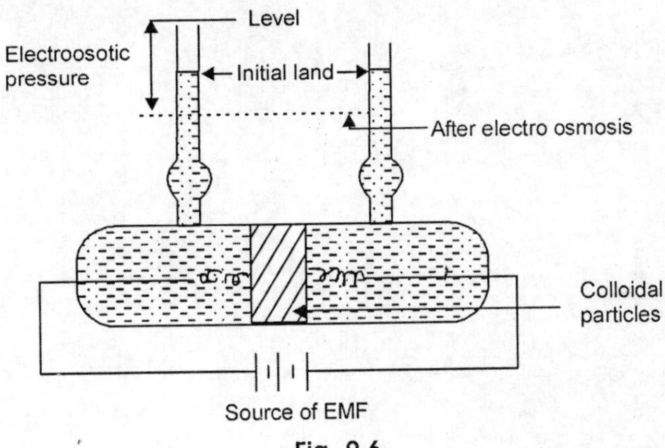

Fig. 9.6

Here, the colloidal solution is sandwitched between the dispersion medium. The level of the liquid in vertical glass tube is noted. An electric potential difference is now applied with the help of two electrodes. It is observed that level of the solvent rises in one of the tubes due to migration of the solvent. A difference of pressure is steadily build up till the movement of solvent due to electro osmosis is exactly balanced by a flow in the opposite direction due to the developed pressure difference. The difference of pressure thus built up is called Electro osmotic pressure.

From the knowledge of the direction in which the movement of solvent occurs, it is possible to determine the nature of charge on the colloidal particles. Flow towards the cathode indicate that the colloidal particles are negatively charged whereas flow in the opposite direction shows that particles are +vely charged.

PROBLEMS

1. Explain electrical double layer and why a colloidal solution contains electrically charged particles?
2. Describe the origin of charge on colloidal particles. What is meant by electrical double layer and zeta potential?
3. Differentiate between:
   (*i*) Hydrophilic and hydrophobic sols
   (*ii*) True solution and colloidal solution.
4. Explain the term electrophoresis and electro osmosis and give their importance.
5. What are protective colloids? Explain how a lyophilic colloid can stabilise a lyophobic colloid. What is meant by the term Gold Number?
6. Write short notes on:
   (*i*) Brownian movement
   (*ii*) Tyndall effect

CHAPTER

10

CERAMICS

10.1 INTRODUCTION

The word ceramic is derived from the Greek word keromikose. The term covers inorganic non-metallic materials whose formation is due to the action of heat.

The American Society for Testing and Materials (ASTM) defines a ceramic article as "an article having a glazed or unglazed body of crystalline or partly crystalline structure, or of glass, which body is produced from essentially inorganic, non-metallic substances and either is formed from a molten mass which solidifies on cooling, or is formed and simultaneously or subsequently matured by the action of heat."

Thus, we can say that ceramics are inorganic, non-metallic materials that can be made into a paste and shaped at normal temperatures. The Shape is then burnt at a high temperature to produce strength in it either by sintering the crystallites together. Some metallic oxides, carbides, silicates and their combination are processed in the same way to obtain different ceramics.

10.2 TYPES OF CERAMIC MATERIALS

Ceramic materials and wares can be divided into following five categories:

10.2.1 Structural Ceramics

Articles used in construction of buildings and other structures belongs to this group, *e.g.*, common bricks, brick block, hollow tiles, roof tiles, drain tiles, ceramic slab for floors, sewer pipe, clinker bricks etc.

10.2.2 Facing Material

Articles used mainly for external and internal facing of building and structure *e.g.*, facing bricks, slabs and oven tiles.

10.2.3 Refractories

Materials which retain their mechanical properties at 1000°C or high temperatures.

These are also used in making various parts of industrial furnaces, ovens and apparatus used at high temperature.

10.2.4 Fine Ceramics

Glazed pottery and porcelain wares are included in this category. Following are the examples of fine ceramics:

Domestic purposes: Wash basin, sinks, dishes, decorative articles, etc.

Electrical purposes: Electrotechnical porcelain.

Laboratory purposes: Chemical wares and apparatus.

10.2.5 Special Ceramics

A group of articles with specific properties utilized in aviation, radio industry, instruments manufacture etc.

10.3 CLASSIFICATION OF CERAMICS

Ceramics may be classified into following two categories:

10.3.1 Non-Crystalline Ceramics

Non-crystalline ceramics, being glasses, tend to be formed from melts. The glass is shaped when either fully molten, by casting or when in a state of toffee like viscosity, by methods such as blowing to a mold. If later heat treatments cause this class to become partly crystalline, the resulting material is known as glass ceramic.

10.3.2 Crystalline Ceramics

Crystalline ceramics materials are not amenable to a great range of processing. Crystalline ceramics may be manufactured by following two methods:

In Situ Manufacturing

In this method ceramics can be prepared in the desired shape by reaction in situ or by "forming" powders in the desired shape and then sintering to form a solid body. The most common use of this method is in the production of cement and concrete. Here, the dehydrated powders are mixed with water. This starts hydration reactions, which result in long, interlocking crystals forming around the aggregates. After sometimes it appears in the form of solid ceramic.

Sintering based Methods

Once a 'green body' meaning roughly held together object is made, it is baked in a kiln, where diffusion processes cause the green body to shrink. The pores in the object close up, resulting in a denser, stronger product. The firing is done at a temperature below the melting point of ceramic.

Sometimes a slurry can be used in place of a powder, and then cast into a desired shape, dried and then sintered. Traditionally, pottery is done with this method.

10.4 PROPERTIES OF CERAMICS

Some important properties of ceramics are as given below:

10.4.1 Mechanical Properties

Ceramics are brittle solids which are very resistant to compression. The strength of ceramics is mainly controlled by following factors; temperature, composition, surface conditions, micro structure, shape and size.

10.4.2 Electrical Properties

Oxide ceramics are generally bad conductors or insulators in their normal oxidation states. The non-oxide ceramics are being used as semi-conductors.

10.4.3 Magnetic Properties

Ceramics containing transition metal ions have shown magnetic properties, because of spins associated with unpaired electrons. Magnetic ceramics are those which contain molecules with odd number of electrons and incompletely filled $3d$, $4f$ and $5f$ orbitals. Magnetic ceramics are bad conductors of electricity. Magnetic ceramics respond to magnetic field, known as magnetic susceptibility.

10.4.4 Optical Properties

Ceramics are transparent as well as opaque. The translucency or opacity in a ceramic body is due to scattering of light resulting due to difference in the refraction at the boundaries of a polyphase polycrystalline material. The optical properties of ceramics depend upon its composition and crystal structure, because extent of polarisation mainly depends upon or controlled by ionic size bonding energy and crystallographic direction.

10.4.5 Physical and Chemical Properties

The components present in the ceramics, such as oxides, carbides, etc., give high chemical stability to ceramics. Ceramics are resistant to highly oxidising and reducing atmospheres and also fluctuations to temperature compactness of crystal structure, high directional character of chemical bonding and high field strength of small cations of high charge are responsible for the stability of ceramics.

10.5 WHITE POTTERY

The term white pottery or white wares usually refers to glazed or unglazed ceramic materials which have white or pale cream color products after firing. White pottery are usually prepared by the mixture of water and china clay (free from iron), feldspar and quartz or flint in suitable quantity. Clay and water mixed together to form a paste and subjected to produce the desired shape of articles. This is followed by careful drying and then firing at 1350–500°C.

10.5.1 Manufacturing of White Pottery

The materials required for manufacturing of whitewares are; china clay, feldspar ($KAlSiO_8$) and flint (SiO_2) free from any compounds of iron. The manufacture of white wares consists of the following steps:

Preparation of Slip

China clay, feldspar, flint etc., are mixed in a suitable proportion with water and stirred to get a cream like paste known as slip. Then other unwanted materials are being filtered from slip.

Filtering and Aging

The filtered slip is passed through the filter paper to remove moisture and then stored for few weaks called aging of slip. This process increases the plasticity of the clay.

Shaping of the Articles

After filtering and aging, the shaping of article is done either by hand moulding or by die pressing. The articles so produced is dried slowly and fired to the desired density in a biscuit over to get porous ware called bisque.

Firing

The dried articles are placed in a furnace. During the process of firing following changes occurs:

- Water is completely removed at 150–600°C.
- Lime stone gets decomposed ($CaCO_3 \rightarrow CaO + CO_2$) at 800–900°C.
- Oxidation of ferrous salt and organic matter occurs at 350–900°C.
- Combination of metallic oxide with silica occurs at 1200–1300°C.

Glazing

To achieve the ceramic surface smooth, non-porous, non-permeable, hard attractive and easy to clean, a mixture of glass forming materials in suitable composition such as lead silicates, borosilicates etc., are applied over the articles.

Decoration

In order to decorate the article, the desired design may be painted on the body before glazing (called under glazed) or the design may be painted upon the glaze followed by another firing to melt the pigment into the glaze (called over glazed).

10.5.2 Properties and Uses of White Pottery

White wares possesses good strength, very low porosity and good translucency. White wares are used as electric insulators, laboratory equipments, crucible, furnaces, dishes, domestic utensiles as well as in aviation.

10.6 GLAZING

Glaze is a glassy coating applied to pottery, the primary purpose of which include decoration and protection. Glaze is a mixture of glass forming materials *viz.*, SiO_2, Al_2O_3, lead oxide, ceric oxide, borosilicates etc. The glaze must contain atleast one alkaline earth metal and should be free from iron and other coloring pigments. For color glazing, colored metal oxides are mixed in proper portion.

Glazing is mainly done for the following requirements:

- To increase the durability of ceramic materials.
- To provide smooth and shining surface to treated materials.
- In order to make the earthware surface impervious to liquids.
- To produce colored and decorative effect.
- To protect the surface from corrosive action of the environment.

10.6.1 Methods of Glazing

Following method are being adopted for glazing: immersion, pouring, spraying, dusting volatilisation etc. All these methods can be usually categorised into following two categories:

(*i*) Salt glazing (solid form)

(*ii*) Liquid glazing (liquid form)

Salt Glazing

In this process salt of KCl or NaCl is poured into furnace at 700–900°C, where the salt volatilised and react with silica (articles) to form glossy or shining and impervious film of sodium potassium silicate.

Liquid Glazing

Liquid glazing is some how superior than salt glazing. In this method fine powder of glaze mixture with suitable proportion of coloring pigments are mixed with water through revolving blunger to get a cream like paste called slip. The articles to be glazed are put in the furnace at 350–500°C. These hot articles are then taken out and dipped into the glaze slip so that glaze materials will occupy the pores in the articles. After this the articles are again fired in furnace at 700–900°C. Due to the firing glaze materials fuse and forms thin glossy films over them.

10.7 APPLICATIONS OF CERAMICS

Ceramics are used for various domestic as well as industrial purposes, *e.g.*,

- Used in making bricks, pipes, floor and roof tiles.
- Used as refractories such as kiln lining gas fire radiants, steel and glass making crucibles.
- Used as white wares, including table wares, wall tiles, decorative art objects and sanitary wares.

- Besides above applications ceramics are being also used in making tiles for space shuttle program, gas burner nozzles, ballistic protection, nuclear fuel uranium oxide pellets, biomedical implants, jet engine turbine blades and missile nose cones.

PROBLEMS

1. Define the term white pottery.
2. Explain the manufacturing of white pottery.
3. Define the term ceramics.
4. Give the applications of white pottery.
5. Write short note on glazing.
6. What are ceramics and how are they classified?
7. Explain the liquid glazing as well as salt glazing of ceramics.
8. What are the important raw materials for the manufacture of white ware products? How are white wares prepared?
9. Describe the various steps in the manufacture of white wares. Discuss the various characteristics of ceramic white ware products and their industrial applications.
10. Write an informative note on following:

    (*i*) Glazes (*ii*) Firing of ceramic wares (*iii*) Feld spars (*iv*) Slip
11. What is meant by plasticity of clay?
12. Write note on glazing of pottery products. What is the purpose of glazing?

INDEX